Woven Together

Woven Together

Faith and Justice for the Earth and the Poor

JAMES S. MASTALER

Foreword by Holmes Rolston III

CASCADE *Books* • Eugene, Oregon

WOVEN TOGETHER
Faith and Justice for the Earth and the Poor

Copyright © 2019 James S. Mastaler. All rights reserved. Except for brief quotations in critical publications or reviews, no part of this book may be reproduced in any manner without prior written permission from the publisher. Write: Permissions, Wipf and Stock Publishers, 199 W. 8th Ave., Suite 3, Eugene, OR 97401.

Cascade Publications
An Imprint of Wipf and Stock Publishers
199 W. 8th Ave., Suite 3
Eugene, OR 97401

www.wipfandstock.com

PAPERBACK ISBN: 978-1-5326-6167-9
HARDCOVER ISBN: 978-1-5326-6168-6
EBOOK ISBN: 978-1-5326-6169-3

Cataloguing-in-Publication data:

Names: Mastaler, James S.

Title: Woven together : faith and justice for the earth and the poor / James S. Mastaler.

Description: Eugene, OR: Pickwick Publications, 2018 | Series: if applicable | Includes bibliographical references and index.

Identifiers: ISBN 978-1-5326-6167-9 (paperback) | ISBN 978-1-5326-6168-6 (hardcover) | ISBN X978-1-5326-6169-3 (ebook)

Subjects: LCSH: Human ecology Religious aspects Christianity | Social justice Religious aspects Christianity | Religion Christian life Social Issues |

Classification: BT695.5 .M39 2019 (paperback) | BT695.5 2019 (ebook)

Manufactured in the U.S.A.   04/15/19

Scripture quotations are from New Revised Standard Version Bible, copyright © 1989 National Council of the Churches of Christ in the United States of America. Used by permission. All rights reserved worldwide.

For
Hawa

"Our life is a faint tracing on the surface of mystery, like the idle curved tunnels of leaf miners on the face of a leaf. We must somehow take a wider view, look at the whole landscape, really see it, and describe what's going on here. Then we can at least wail the right question into the swaddling band of darkness, or, if it comes to that, choir the proper praise."

Annie Dillard
Pilgrim at Tinker Creek

Contents

Foreword by Holmes Rolston III | ix

Acknowledgments | xiii

Introduction | xv

1 We Need New Stories | 1

2 Facing the World as It Is | 22

3 A Faint Tracing on the Surface of Mystery | 44

4 Into the Darkness with Hearts Ablaze | 77

Bibliography | 109

Index | 135

Foreword
Weaving What Together?

BY HOLMES ROLSTON III

You have in your hand an intense effort to weave things together, although Mastaler will instantly insist that the weaving is of words and life. This book in hand is pointless unless it points you toward living in a "more just and verdant world" (94). But that requires interweaving multiple dimensions of what we as individuals and as communities think and do. That requires "the creative potential of powerful storytelling" (3). Perhaps a foreword can help if it considers what is woven together in this powerful story.

What is mainly woven together is Christian concern for Earth and for the poor on Earth, environmental ethics, and environmental justice. "If we first acknowledge the indivisible necessity of both economic development and ecological conservation for shared planetary flourishing, then we have our twin pillars of sustainable development" (36).

Two other themes interwoven here are the theoretical and the practical. In an analysis of medical care in developing nations, we encounter a little girl in rural Bangladesh named Hawa, with failing kidneys as a result of severe malnourishment and dehydration. Trying to distinguish the different degrees and causes of poverty, we decide to ask a poor, disabled woman in Zimbabwe to define poverty: "What kind of definition of poverty do you expect me to give you, which is better than what you have seen with your naked eyes" (28–29)?

Two further interwoven themes are the past and the present, from ancient Genesis to Pope Francis' recent encyclical letter, *Laudato si'*. The pope himself sets an example for us, drawing on the past, adapting it for the present, and with an eye for the future (51). We visit Augustine of Hippo and St. Thomas. One might think that the ancient seers, whatever their wisdom, have little to say to help us with global warming. They didn't even know they were on a planet. "Many of the old stories simply do not incorporate images of the world as we now know it, and they do not cultivate the kind of moral concern for other species, and the planetary life support systems on which we all depend. Critically and creatively imagining new stories is an essential way to usher in a dramatic shift in the predominating collective consciousness" (74–75).

But maybe what those who knew the law and the prophets did know about human nature, our self-interests, our selfishness, our short-sightedness is a key to solving this new unprecedented contemporary problem of curbing global warming. They fully recognized how humans turn in on themselves, rather than seek the common good. Mastaler can be an insightful guide in an effort to help the world "shift from a generally and overly anthropocentric approach to a responsibly retrieved and yet newly constructed creation-centered framework more capable of supporting planetary flourishing" (75).

Mastaler is challenged to interweave human uniqueness and human embeddedness in nature, humans fitted into the ecosystems they inhabit. "Human continuity and participation in the larger universe story frame perceptions of the human person with special or distinctive attributes as simultaneously wound up with and connected via a shared evolutionary story to every other being with which our species shares the planet" (73). Humans and nature are not polar opposites. Replace divergence with convergence. Figure out how you are "distinctive" and "simultaneously wound up," integrated into everything else. See if Mastaler helps. "I wish more of us could approach the world with the faith of a peony" (20).

He seeks "enfleshed spirituality" (66). Classically, we might have called this the human sensing of simultaneous transcendence and immanence. The contemporary challenge is to blend the spectacular new powers gained in science and technology, distinctive to *Homo sapiens*, with wisdom about using and constraining these powers. "As our collective human power and presence on the Earth grows, the need for careful, judicious, and humble wielding of this newfound power also increases" (98). Enfleshed and embedded, as are the other animal species, yet we are considering geoengineering the planet. Weave those disparate dimensions of human nature together! Get yourself an "earthy sacramentalism" (62).

This book in your hand interweaves the discursive and the pointed, surveying what has been said by dozens of others, often ending with an aphorism to provoke you to further thought. "Cultivate a gritty kind of faith" (96)! The book reconsiders accounts of incarnation and redemption across the centuries of Christian theology, working out this enfleshed spirituality. The sum of it: "There is no salvation from the world, only salvation with the world" (71). The book recalls and surveys cosmologies, ancient, medieval, contemporary, East and West. Then: "We are the cosmos writ small" (70).

We find ourselves wondering whether and how we can interweave rights, values, instrumental values, intrinsic values, goods, goods-of-their-own, legal standing, natural law, moral law, civic law, ownership, animals, sentience, persons, respect, reverence. Mastaler tries to put everything together. The weaving together of these ideas will take some rather sharp distinctions, or else it will just be a muddling together, bringing as much confusion as clarity. Readers will watch for any use of weasel words where different parties to the conversation seem to agree because they use the same word; but, since they mean different things by this same word, the apparent agreement is superficial. "Sustain" and related words occur some sixty times in the text. "Sustainability" - everybody wants it - but sustainable what? Sustainable growth? Development? Profits? Resources? Biosphere? The devil is in the details. Mastaler chases these interwoven details, and invites you to come along.

A particularly devilish problem is how to weave together "myth" and "truth." Scholars have found this perennially challenging. Mastaler calls for "a kind of myth-making on an epic scale, which is to say, one that is up to the equally epic task of ushering in 'a period when humans would be present to the planet in a mutually beneficial manner.' New myths are needed that are capable of penetrating the deepest levels of human consciousness. They are needed to inspire and motivate changes to deeply held assumptions one holds about oneself in relation to others and the world" (72).

Of course, if this adventure is simply choosing what new or old myths we prefer, what spin we choose to put on ideas we have inherited or experiences we have had, then my myth is as good as yours, and yours as good as mine. We need some way of testing better and worse myths, which, hopefully, will better fit what we consider to be facts of the matter—about

Earth, a planet in space, about the biosphere and what threats it faces, about human nature and decisionmaking. So watch how Mastaler mixes truth and myth, with his invitation "into the darkness with hearts ablaze" (77). You are going to be challenged to think big, to weave yourself into the Earth, into the cosmos. "Our consciousness, our curiosity, and our sense of wonder and awe are gifts the universe has brought into being through us, and we can use them to advance the flourishing of life across the cosmos" (73). Maybe you will wonder whether that is too grandiose. Humans are the wonder of wonders on wonderland Earth, but their influence is quite limited in a cosmos that is 93 billion light years across. But you will not wonder about Mastaler's hyperimmense vision.

Here, in his conclusion, is the hoped-for interweaving:

> I so desperately wish more of us would see and engage the world in this way—to more clearly and more often experience that deeply mysterious and seductive sense of joy and peace that really is all around us. We only have to look for it. Seek God in all things, through the microscope, the telescope, and everywhere in between. The same energy that brought the cosmos into being, sustains it still; it fills your lungs, it courses through your veins, and it animates your spirit. There is hope in that. (107)

Acknowledgments

It is my great pleasure to acknowledge the many folks who were instrumental in helping to make this book possible. First, my colleagues at Loyola University Chicago, especially my mentor William C. French for his willingness to help me work through the earliest iterations of the ideas presented in this book. I am thankful to Michael Shuck, Robert Ludwig, and Nancy Tuchman for their ongoing support and encouragement along the way. Mary Elsbernd and Mary Evelyn Tucker have inspired me to believe this kind of work is worth coming into being. I am indebted to Holmes Rolston III for his helpful comments on later drafts and to my friend, Julie Meadows, who helped me to find my writing voice and taught me how to more clearly articulate it. I am grateful to Richard McCarty for his sage advice and kind ear, as well as to Ryan Hoffman and Paul Ott for so generously agreeing to read and offer comments on later drafts. Finally, thank you Michelle, Michael, Jonathan, and Bobbi Lynn whose stories are etched in here alongside my own. Sherry and Robert, Carol and Jerry—thank you for cheering me onward. Jason, thank you for believing in me.

Introduction

I HAVE ALWAYS BEEN deeply moved by encounters in the natural world. It is where I have felt most at home and most myself. The church is a close second, but my relationship with it has been a bit more complicated. After a decade of graduate study in Christian ethics and theology, and several years of teaching and discussing these topics with students, that relationship is still complicated. I do not think that is a bad thing. The primary subjects in this book—faith, equity, and ecology—are related to each other in a somewhat complicated way. The way I came to see their connections was by reflecting on the transformative experiences shaping my own life. These include early experiences with religion and nature, as well as experiences of poverty, pollution, and overwhelming human need. They inform the point of view I bring to these subjects.

Religion was something of a contested space in our house while growing up. My dad was Roman Catholic and mom was Protestant. They disagreed vehemently on how to live out various aspects of their faith, but they loved each other and learned to prioritize that over everything. Perhaps that is why we did not regularly attend church at the time I was first learning about Christian faith. My grandmother—a Roman Catholic-born convert to Lutheranism and then evangelicalism—faithfully attended her nondenominational church each and every week and I sometimes joined her for services. So, my earliest memories of church were in her small, Bible-believing, evangelical community where I was dedicated to Christ as an infant, and accepted him as my personal Lord and Savior before I even knew how to read.

Some of my earliest memories involved my grandmother quizzing my sister and me on the different books of the Bible. I internalized those stories so that their main characters became a part of my everyday life. Daily morning devotions were a habit I grew into and practiced for some time as a young adult and even later after I left home for college. I would pore over Scriptures while eating my Cheerios. Reading the Bible in a year was a thing—I did it several times. I began and ended every day in prayer under her tutelage. I am thankful to have been exposed to these perspectives in and on Christian faith from such an early age.

As I grew in my understanding of faith I longed for a version of it that felt more authentically rooted in what I knew to be true of the world and not merely what others told me to believe was true. When I look back on my encounters with the church, what I appreciate most are the moments when folks in the church were willing to meet me where I was and as I was. This includes their willingness to join me in grappling with incredibly complex questions, rather than demand I settle for easy, unsatisfying answers. There was a willingness not only to *let* me reach out for something none of us will ever be able to fully get our arms around, but a sense of shared joy in our common curiosity to explore the wonder, awe, and even fear we sometimes feel in the face of deep mystery. Open to the unknown but grounded in living experience—that's my kind of faith.

Encounters with poverty, both at home and around the world, are part of that living experience. Though I did not then think of my family as poor, my sister and I would go with my mom to the food pantry on weekends to pick up groceries donated by the local community. Our weekly haul often included powdered milk in boxes instead of the kind that came in a jug. When reconstituted, it makes for a nice toast and gravy dinner—high in carbs with some protein but otherwise not very nutritious. Usually, those packages, like our peanut butter, would come in generic containers with black and white labels, the moniker for surplus foods packaged by the USDA for food pantries.

I distinctly remember one strange Christmas when I received purple gloves one size too big and a pair of fancy headphones intended to be used with a Walkman, only I did not have a Walkman. It turns out that year we were one of those families with our name on an ornament in the parish hall that other families buy gifts for at Christmas time. I cannot help but believe it was because of the generosity of those other families at my church, and the generosity of the community in our poor but safe suburban enclave, government programs like Women Infants and Children (WIC), and the support of my family's strong social network that I am lucky enough to say things ended up ok for me.

Introduction

Those supports freed my parents to focus on their work, career development, and professional advancement. By the time I was in high school, our large family of seven was living free of government assistance and the charity of our church family. We were able to move into a lovely old home with bedrooms for all of us and we had a spacious yard to boot. I planted some peonies in the back. I nurtured my garden and found it nurtured me too. Working with the soil, the sun, birds, and bees became something of a spiritual practice—a practice and a process of mending not only my own brokenness but also my broken relationship with the Earth and those with whom I share it. While I did not know it then, I now see it was an influential part of my heart's ongoing love affair with all life on Earth in its many and various forms.

The inward and outward journey begun in my garden took a few more turns when I went off to college. My grandmother was the youngest of ten in an Italian Catholic immigrant family in Chicago. She believed it was her responsibility to stay home and take care of her ill mother, and that meant quitting high school to do so. My grandmother went on to have eleven of her own children; my mom was one of them and my dad was one of ten. Self-sacrifice for the needs of the family was the atmosphere I was raised in, with several generations worth of that ethic steeping into our collective consciousness—it was how poor, big families survived.

I guess I was selfish enough to stake claim to my own path. Privileged enough, too. I consider myself quite lucky to have gone on to college. I know now that even when first-gen students like myself get into college they often struggle. I not only completed my degree program, but I learned how to flourish in my new context. North Park University is where I really came alive as a person—where I eagerly sought out every opportunity the generous faculty made available to me, and where I grew into the self-aware, independently minded, and others-focused young adult all great universities aspire to form. I do not intend to brag—it has taken me a long time to feel comfortable celebrating my achievements. I am still learning.

During my time at North Park, I spent a summer studying in China. It was then my outward journey took me across the globe but also where I made some strides on my inward journey too—a pilgrimage of self-reflection. My friends and I went hiking in the mountains one afternoon. We were far enough into our trip for the shiny, thrilling newness of being in China to have begun to dull and for the strange differentness of the place to begin grating on our nerves. Our return home was still several weeks away.

Introduction

I was feeling frustrated by my inability to learn Chinese and communicate my ideas with my Chinese classmates, the lack of privacy that comes with communal living (and showering), and I started longing for more familiar food (especially peanut butter). So, the daylong hike up the mountain came during a period when I needed some time to reflect. Until the week before that hike, I felt I had been living the experience of a lifetime. But, I began to wonder what it all might be for.

I arrived at the summit, hot, sweaty, and nearly exhausted, and then found an isolated ledge where I could see for miles. There, alone with my thoughts, I was able to breathe in the cool mountain air and really notice the bright sun. I felt a kind of calm and wholeness that weirdly seemed to reconcile the disparities I was struggling to make sense of at the time. After several deep, calm breaths I exhaled a prayer that was partly a pledge, part mantra, and part supplication: "Please, never allow me to forget that so much joy and beauty exists in this world. Let me find some way to access it, and to remember this moment and this feeling once I climb down from this mountain."

As this was emerging in my heart and mind, a butterfly gently fluttered around me for what seemed like an eternity inside an instant before landing on my right thigh. It sat there as I sat there, both of us seemingly out of place in that arid context, as though the two of us were the only creatures in a vast lonely world. Its presence comforted me and I felt so dramatically un-alone, whole, healed, and fully present. Then the butterfly fluttered along and I was back on the mountain surveying the horizon, alone with my thoughts once more. The moment had passed, yet that experience remains one of the most sacred moments in all my life.

After North Park, I had planned to join the Peace Corps and came close to a post in their environmental education and conservation program. As a requisite for placement, however, the federal government required extraction of my wisdom teeth before they were willing to send me to a remote corner of the globe where my teeth might become impacted and cause a medical emergency. I could not afford the surgery, so I scrapped that plan. My search for other options eventually led me to my church's global missions office. I liked their model of accompaniment in which a member from one congregation is sent to a partner congregation to learn from and work alongside the local community's ongoing development projects. This is opposed to evangelization models focused on spreading word of the gospel. It seemed more "walk the talk" and I often see more value in deeds than

Introduction

words. My church body was willing to send me to Bangladesh for several months of language training and then on to a longer-term placement at a small, rural women's maternal health clinic and community development center.

I saw a lot of things in Bangladesh that challenged the generally black-and-white way I had sometimes perceived the world. One day I went for a walk to get out of my head and take a break from my Bangla language studies. As I was crossing the street, a thin and clearly unbathed woman carrying her obviously undernourished child approached me, hoping I might spare some change, a few *Taka*. I knew enough Bangla to understand her but I felt perplexed as to how I should respond. I had heard that local pimps would send women out with drugged babies in order to ask foreigners for money. Should I avert my eyes and keep going? Should I risk inadvertently supporting her and her child's exploitation? Encounters like this happened nearly every time I left my apartment but something about this one moved me. I simply could not keep walking.

I decided I would not give her money and offered to try to buy her some bread. When she told me she really needed "dudh" or milk for her baby, I became hopeful that this situation was not the one I had been warned about and instead I became worried for the child. We walked together for some distance to buy the milk. I wondered what on earth I was doing. I knew I could not do this every single time I was asked. At best it would buy the baby a few more days. I knew I was not fixing any real problem.

The milk passed from my hands to hers. She smiled in gratitude. I went home and cried. Actually, I wept. I wept because I could not stand the thought of so many people walking past this woman and her child. I wept because I did not know where her child's next meal would come from. I wept because there are so many people in this world who are hungry and need so very little just to make it through one more day.

The experience reminded me of an admonition hung within the sanctuary of my congregation in Galesburg, Illinois. It was a giant banner with a conch shell on it, hand-crafted by ladies in the sewing circle with words cut from silk that read: "blessed to be a blessing." What does a phrase like this mean in a situation like the one I encountered in Bangladesh? I do not think I was much of a blessing, and I certainly did not feel blessed. I did not feel blessed because I knew I did so very little amidst a sea of such great suffering. I was so often completely overwhelmed by my own inadequacy in the face of such great despair. Looking at despair directly in the face is

Introduction

harder than looking away. What kind of global society have we created in which so few have so much when so many have so little?

After Bangladesh, I entered a graduate program in social justice. I went to East Africa for a few weeks to learn from local community leaders who are working to transition their communities out of poverty. I was moved by the disparity and degradation I saw in the slums of Kibera and Mathare. The nearly overwhelming poverty and the environmental degradation—the squalor of the living conditions—appeared to be two sides of the same coin. The connections between social issues like a lack of access to clean drinking water and environmental issues such as polluted waterways were suddenly quite striking and clear. People cannot grow food that is safe to eat on land contaminated by raw sewage or industrial waste—or on land that is as bouncy when you walk on it as a children's rubber-coated playground because it is covered in a layer of plastic bags several inches deep. The Kibera and Mathare I saw on my visit were places where the local environment is so severely degraded by human poverty and overwhelmed by human presence that it struggles to support any kind of life at all. We make the land incapable of nurturing the kind of healthy human flourishing every person deserves.

When I look at the world, I now see it through two lenses that affect the way I read just about everything. I see the world in terms of the ecological systems in which people are embedded and in which our lives are woven together with the lives of others. I also see the world in terms of its social structures and institutional processes—the way our lives shape and are shaped by our interactions with others and the communities we build. In Kibera and Mathare, what was once a bountiful land could no longer support human flourishing because so many different kinds of relationships had become broken.

The background structure of this book mirrors my own journey of faith: I have come to understand faith and action as inextricably connected, where faith is always to be animated by action. So, chapter 1 makes the case that religious stories are a central part of the ideas about the world that matter because they affect how we act. Chapter 2 describes the world as it is, grounding all the conversations that follow in the lived experiences of those at the margins of society. The point is that these realities must inform the conversations about God and the world that people of faith are having in the church. Chapter 3 then turns to Christian religious stories to clarify which stories are helpful in speaking to these experiences, which

Introduction

stories might become helpful if they are rescripted, and which stories probably need to be trimmed away. Finally, chapter 4 returns our focus to the way stories of faith about how to live in the world connect to how people can actually live their lives. It offers a theology of mobilization, calling for Christians to live out a gritty kind of faith that arouses a deep love for the world and works to make justice happen.

All people deserve a chance to pursue the things they need in life. We all need the basics of food, shelter, and clean water. We also need things like medical care when sick, access to educational opportunities appropriate to our abilities, social networks of care and compassion, as well as space for reflective contemplation. There are so many broken relationships among people and between people and the planet that it is hard to imagine we will ever bridge the divides and heal those relationships. We cannot let that stop us from trying. Christian faith, when it is animated by action and sustained by hope, is more than a personal devotion to Christ because it is a commitment to a way of life lived in his footsteps. It is my hope that you may find something in these pages that gets stuck in your craw—something that challenges the way you see yourself and the world—something that does not rest easily, but something you are willing to wrestle with for a time. I hope the ideas that follow compel you to do something, to make something, to be a part of the movement toward a more just and verdant world.

1

We Need New Stories

I HAD SO MUCH hope. The year after former US Vice President and climate activist Al Gore's *An Inconvenient Truth* made its theatrical debut, I started a graduate internship and then accepted an advocacy position with the Illinois Chapter of the Sierra Club.[1] The Club was abuzz with staffers celebrating the "game-changing" nature of the film and its effects on public awareness about climate change. The film contained persuasive graphs and charts. The science presented was solidly vetted and held up to scrutiny. It was narrated by an influential political figure. It won two Academy Awards, and Gore went on to share a Nobel Peace Prize with the Intergovernmental Panel on Climate Change for "informing the world of the dangers posed by climate change."[2] So many people felt invigorated by what seemed newly possible concerning climate policy at home and abroad. Things were looking up! The excitement was palpable.

The optimism was short-lived. The United States did not soon take any concrete political action to mitigate the climate problem. Few would have believed that any international agreement behind which the US would throw its support was still a decade away from that point. Climate concerns among the US public quickly dissipated. Climate deniers countered with a slew of challenges and counter-arguments, albeit baseless. Gore's graphs

---

1. Gore, *An Inconvenient Truth*. Some of the ideas in this chapter were published in an earlier iteration. See Mastaler, "Role of Christian Ethics," 43–48.

2. Nobel Foundation, "The Nobel Peace Prize 2007." http://www.nobelprize.org/nobel_prizes/peace/laureates/2007/gore-bio.html (accessed September 4, 2018). See also Grandin, ed., *Nobel Prize*.

and charts were soon all but forgotten, lost to a mist of doubt and confusion. What at first appeared to be a catalyst for the climate movement quickly faded. This was my first foray into organizing around something that mattered, and I failed. I was defeated.

What happened? Convincing people that climate change is a serious problem is quite a bit more complicated than many of us ever imagined. We live in an era of the twenty-four–hour news cycle, and several networks include entertaining, lively debates as a part of their allegedly fair and balanced coverage of controversial issues. Sometimes the networks serve one entrenched special interest or another by intentionally skewing these debates with the questions that are asked or by whom they choose to include in the debates. At other times the issue is unintentionally distorted when well-meaning journalists give undue attention to climate deniers by allowing such a small number of people holding a given perspective to be equally represented in a debate—an overwhelming 97–98 percent of climate scientists agree on the anthropogenic aspect of climate change.[3] When the issue is presented as though experts in the field give equal credence to each perspective, while the "debated" perspectives do not, in fact, hold equal weight, the media does the public an incredible disservice.[4] The general public rightly perceives media coverage on climate change as the confusing mess that it has been.

The environmental community's primary response to this confusion has been its long-standing attempt to improve scientific literacy among the general public. Their focus has been laser-like: more graphs, more charts, better arguments to call out the false science. Outreach and messaging have been almost exclusively geared toward confronting climate denial head-on. For all our effort, this has not worked. Scientific knowledge alone simply does not generate the kind of knowledge capable of moving the general public to take action on important environmental problems such as climate change. What could we have done differently?

By focusing on improved scientific literacy rather than the role of deeply held values and beliefs and in building bridges and coalitions with faith-based religious communities, we in the environmental community missed a valuable opportunity. Stronger bridges would have better directed the movement toward the kind of conversations able to more powerfully

---

3. Anderegg et al., "Credibility in Climate Change," 12107–12109.

4. Gore, *Earth in the Balance*, 38–39; See also McKibben, *Age of Missing Information*; Stauber and Rampton, *Toxic Sludge*.

sustain enduring public engagement for the long haul. Directly emphasizing the scientific reality of climate change to deniers is not as effective as framing it in terms of the sacred stories and religious narratives—the worldview-level of ideas—that shape a person's fundamental beliefs about who they are and what kind of people they think they ought to be in the world.[5] The movement failed to generate substantial public concern for its issues or create any lasting change. The creative potential of powerful storytelling was not, and has not yet been, fully tapped.

Scientific information is filtered through the lens of a person's worldview—a person's general way of seeing the world that makes them more or less receptive to certain ideas. We now know that the more scientific knowledge an individual has, the more likely that individual is to use that knowledge to affirm *preexisting* values and beliefs about the way they think the world works. In one important and groundbreaking study, the authors marvel at how well "equipped ordinary individuals are [at discerning] which stances towards scientific information secure their personal interests" and they posit that "the reward for acquiring greater scientific knowledge and more reliable technical-reasoning capacities is a greater facility to discover and use—or explain away—evidence relating to their groups' positions."[6] The study found that "[m]embers of the public with the highest degrees of science literacy and technical reasoning capacity were not the most concerned about climate change. Rather, they were [those] among whom cultural polarization was greatest."[7] In other words, the primary distinguishing factor regarding whether people are likely to accept or deny climate change has less to do with their comprehension of the science and more to do with the cultural and political ideas they use to interpret it. The study concluded, "cultural world-views explain more variance [in beliefs about climate change] than science literacy [. . .]."[8] So, the scientific community and environmental advocates can confront climate deniers head-on with all the facts they can muster. The simple reality is that they are unlikely to change very many minds.

---

5. Bain et al., "Promoting Pro-Environmental Action," 600–603. It should also be remembered that as an observational, evidence-based science, it is not the role of scientific inquiry to make ethical normative claims about the world or its findings.

6. Kahan et al., "Polarizing Impact of Science," 733, 734; See also Lakoff and Johnson, *Metaphors we Live By*; and Lakoff, *Moral Politics*.

7. Kahan et al., "Polarizing Impact of Science," 732–35.

8. Kahan et al., "Polarizing Impact of Science," 732–35.

For many people around the world, religious narratives profoundly shape the way they live in the world. In the West, Christian traditions have held a historic and ongoing role in shaping both individual and public consciousness. Christians, spread across diverse communities of identity with unique histories and traditions, do not represent a uniform, monolithic community. Still, many people of Christian faith see in their traditions and their communities a deep source of meaning that enriches their lives and unites them in solidarity with others of their faith. The clergy play an influential role in helping members of their community to make sense of everyday challenges, whether as a part of pastoral counseling, Sunday sermons, weekly Bible study groups, or increasingly by public commentary in literature, radio, television, online blogs, and podcasts. They do so often by telling powerfully moving stories, relating the deeply held values and beliefs embodied in their respective traditions with the more immediate problems of daily living. The most proficient ministers are often skilled storytellers and cultural translators who are comfortable navigating between the concerns of everyday life and a people's most deeply held values and beliefs.

This is an impressive skill that need not compete with, but rather complement the kind of knowledge scientific discovery offers. Scientists have developed methods to observe the world and draw evidence-based conclusions about it, but religious and spiritual narratives often shape how those conclusions are perceived. US conservatives, especially those initially skeptical of climate science, are "more likely to embrace climate science if it comes from a religious or business leader, who can set the issue in a context of values that differ from those of an environmentalist."[9] I remember when this revelation appeared in an article that was reprinted and passed around climate advocacy circles and LISTSERVs during my work with the Sierra Club. It makes such clear sense of the dissonance between US conservative voters on the one hand, and climate scientists and environmental advocates on the other.

During my time with the Sierra Club, and especially in representing them at United Nations climate conferences, I often encountered confusion and sometimes resistance to my work as bridge-builder between environmental and faith-based groups. Bridge building between these communities is difficult; they are separated by knife-wielding doubt that cuts both ways. For some in the environmental community, my work is seen as a non

---

9. Mooney, "Science of Why," 40–45.

sequitur—they do not picture how bridges could be built between the two communities, or they do not imagine how those bridges might be helpful. Others, especially those who make a pastime of lambasting climate deniers, make a point of reminding me of the counterproductive role evangelical Christians have played in hampering climate negotiations in the US.[10] I often wonder whether my activist colleagues assumed I was merely naïve or whether they simply thought my work was so unlikely to be successful that their best advice was often to suggest I give up the effort and shift strategies. More than once I felt the sting of condescension with the suggestion that touchy-feely religious ideas really had nothing serious to contribute to the environmental movement beyond the greening of congregational buildings and their organizational operations.[11]

As it turns out, the US public's general lack of concern for issues like climate change, or a failure to accept even its existence as a scientific phenomenon, is partly a failure of the imagination. It is a failure to imagine oneself and the larger world in a way that fits into a more comprehensive understanding of the way the world works—different from that understanding shared by one's community of influence, whether it is a religious, social, or political affiliation.[12] It is not so much that people cannot understand the basic science of climate change (as complicated as it is) but rather that the science is generally not approached in a way that can be

---

10. The distinction here between evangelical Christians and all Christians as a whole is mostly mine. I was often surprised by how many people paint with such a broad brush when they talk about "Christians" and "Christianity." The greening of Christian thought on the environment, in streams of US Protestant and Roman Catholic social thought and instruction, is nearly as diverse as are the traditions themselves. For a sampling of that diversity, see American Baptist Churches USA, "Creation and the Covenant," 34–39; Evangelical Lutheran Church in America, "Caring for Creation," 215–22; United States Conference of Catholic Bishops, "Catholic Social Teaching."

11. One study suggests that self-identified Christians report lower levels of environmental concern than do non-Christians. See Clements et al., "Green Christians?," 1–18. Even if some Christian communities have not yet become more green, there is always the potential that they could become so in the future. Still, as an example of the intense skepticism regarding the potential effectiveness of greening Christian thinking, even from within religion and nature circles, note how Dr. Bron Taylor (Founder of the International Society for the Study of Religion, Nature and Culture, ISSRNC, and Professor of Religion and Nature) posted an article on the ISSRNC's Facebook page referencing Clements et al.'s findings and alleging it is "[m]ore evidence that runs against the 'greening of Christianity' case." For both, see Kwok, "Greening of Christianity'?," and Taylor, "Facebook [Group]," Posted July 31, 2013, https://www.facebook.com/groups/ISSRNC (accessed August 1, 2013).

12. Kahan et al., "Polarizing Impact of Science," 732–35.

easily integrated into the deeply set stories operating on both the conscious and subconscious levels in most people's minds. Those stories are usually shaped and affirmed by an individual's community of influence. Whether those stories can help make sense of something like climate change determines in large part how a person responds to it and how they might begin to go about integrating it into their worldview.

If climate scientists, environmental advocates, and their political allies genuinely wish to see a change in direction, then they will need to work with (rather than against) the tremendous creative potential of those who are much more fluent in, and adept at, the language of deeply held values and beliefs. Religious ethicists and religious leaders have a crucial and creative role to play in this, but so too do the billions of people around the world who live and breathe their faith every day. If we are to have any chance for a better future than the one grimly predicted by the world's best science under business as usual scenarios, then we all need to wake up and get serious. We need a rallying cry that actually rallies, inspires, and sustains the hard work ahead of us.

STORIES SHAPE HISTORY

Years before *An Inconvenient Truth*, Gore reflected on the perilous state of the Earth's ecological decline. "The more deeply I search for the roots of the global environmental crisis," he wrote, "the more I am convinced that it is an outer manifestation of an inner crisis that is, for lack of a better word, spiritual."[13] This crisis before the human species emerges from deep and profound issues, not only the technical limitations of our science. Those issues are embedded in the very structure of the collective consciousness of the societies people have built, particularly in the West. Larry Rasmussen argues that the crisis "rests in the alienated way in which we conceive ourselves apart from nature."[14] The ecological crisis stems from a collective disorientation. What is required is a paradigm shift in the way a major segment of the world's people understand themselves in relation to others and the world.[15] We need first to consider the way stories about who we are and

---

13. Gore, *Earth in the Balance*, 12.
14. Rasmussen, *Earth Community, Earth Ethics*, 182.
15. Gore and Rasmussen are far from the only people who approach this challenge in a way that focuses on worldviews and call for a shift in orientation. For various examples of others, see Cobb, *Christian Natural Theology*; Santmire, *Travail of Nature*; Nash, *Loving*

how we think we ought to live in the world have also come to shape the world, as we know it.

There are problems with some of the deeply embedded stories that have come to shape the history of the West, and they are in need of transformation. The most troublesome aspects of those stories comprise what may be appropriately called a modern turn in the general ethos of Western thought. That turn embodies an interrelated series of events each sharing a measure of complicity. It encompasses the division of science and theology, the Industrial Revolution, and the rise of the Market system. For people of Christian faith, it is useful to recognize the way Christian traditions have helped shape, and in turn have been shaped by, those stories that have come to dominate ecologically problematic streams of thought. To manage the dense tangle of vines denoted by such a complicated history of ideas and events, we will first need to untangle them, decide what is worth keeping, and then prune away the rest.

In the pre-modern West, the study of creation was generally considered the purview of theology. Relations among people, relations between God and people, and the power and presence of God's activity in and amidst the world formed three main spheres of theological inquiry and contemplation: people, God, and nature. The natural order of creation factored as one of three prominent themes in major streams of pre-modern Christian theology and ethics.[16] Nature was perceived in some streams of

---

Nature; Oelschlaeger, *Caring for Creation*; Berry, *The Great Work*; Hessel and Ruether, eds., *Christianity and Ecology*; McFague, *Life Abundant*; Schaefer, *Theological Foundations*; Bouma-Prediger, *For the Beauty*.

16. William C. French argues that the premodern context was fundamentally different from the modern in the sense that nonhuman nature would have been perceived as quite vast, relatively stable, and on the whole generally resilient in the face of human presence. Scientific understanding and technology had simply not yet advanced far enough for humanity to use it in a way that might pose a significant threat to the Earth's continued viability. For the vast expanse of human and planetary history, human populations have been relatively small and fragmented. Human life-spans have generally been short and quite vulnerable to diseases and all sorts of plagues—the transmission mechanisms for which were often steeped in superstition. Relatively simple agricultural practices and the utter dependence upon natural systems meant natural disasters and phenomena like drought, extreme weather, and pests were a matter of life and death for most people. Conversely, nature seemed limitless in power and expanse and a sense of awe, mystery, and even fear would have been quite an appropriate response to humanity's fragility before the power manifested in nature. As French notes, "[t]he order of nature seemed to be a given, something whose existence and ongoing presence could be comfortably assumed," so the world would have seemed like "a solid stage upon which human lives danced in our brief course." See French, *With Radical Amazement*, 55.

thought as inherently good by virtue of it being created and loved by God, or as good and deserving of respect for the magnitude of its power in the face of human vulnerability and the special way it would have been believed to reveal the Creator to people.

At the end of the medieval period and along with the rise of modern science toward the beginning of the modern period in the 1500s, we can track the emergence of a profound dualism that influenced trajectories of thought in theological reflection. Jame Schaefer, referring to this period, notes how "reflection on the sacramental character of the physical world waned and the world was reduced to an object for human investigation and exploitation."[17] When the world ceased being a primary revelation for the word of God, it changed the way people of Christian faith came to regard it. "The Book of Nature and the Book of Scripture," as N. Max Wildiers puts it, "which were once seen to proclaim the same wisdom, now appeared to speak a different language and could in any case not readily be reconciled with each other."[18]

The rise of modern science in Western Europe as a distinct field of inquiry separate from theology helped create a dualistic metaphysics in which the realm of nature, or the study of creation, was cleaved from theology and placed within the purview of scientific investigation. This left theology with the spheres of God and people, as science adopted the world of earthly objects and theology was left the heavenly realm of souls and spiritual concern.[19] The individual before God grew in prominence in theological reflection, while concern for earthly matters declined.

From the 1500s through the 1700s, there was a general shift from organic conceptions of the cosmos toward more mechanistic models. Carolyn Merchant calls this a shift to a machine metaphor that she argues came to redefine the way the natural order was perceived.[20] Scientists such as Francis Bacon (1461–1626) and Sir Isaac Newton (1642–1727), as well

---

17. Schaefer, *Theological Foundations*, 86.

18. Wildiers, *Theologian and His Universe*, 81.

19. As the prominence of a strong doctrine of creation began to wane, an emphasis on the human framed within the dynamism of history and in relation to God alone grew. See French, "With Radical Amazement," 59. A subsequent theological and philosophical turn to the subject further minimized any role for the doctrine of creation in streams of both Protestant and Catholic theology in the late nineteenth and twentieth centuries. See French, "With Radical Amazement," 56–58. Also, Gustafson, *Theocentric Perspective*, 82–85.

20. Merchant, *Death of Nature*, 192–93.

as philosophers such as René Descartes (1596–1650), helped develop the machine metaphor as a frame for their approach to the natural world. This view of nature, as a kind of machine, became a new and fundamentally reductionist image of the natural world directly inspired by the impressive technological innovations of the time. It facilitated other aspects of the modern turn, such as the rise of market systems, which in turn was reliant upon the transformation of natural resources into products of production and consumption.[21]

At bottom, the machine metaphor was as much a product of its historical and social context as it was a shaper of it. It was part of a complex matrix of factors that, as James Nash writes, included "population pressures, the development of expansionistic capitalism in the forms of commercialism and industrialization (particularly ship-building, glassworks, iron and copper smelting), [. . .] and the triumph of Francis Bacon's notions of dominion as mastery over nature."[22] Commercialism and industrialization remain an indispensable part of modern market systems today, but they arose within an image of the earth as inexhaustible—a notion we now know to be false. While the immediate pollutive impacts of commerce and industry on the land, water, and air would have been obvious at this time, those driving this movement probably could not have foreseen its globally cumulative impact over the next several generations.[23] The view of the world and our place in it that emerged during this complex confluence of events played an important role in birthing the modern ecological crisis.[24]

The rise of the modern sciences, propelled by the Baconian emphasis on a technological appropriation of nature for human use and the Cartesian machine metaphor as a way to understand nonhuman animality, helped shape the new theologies emerging in the modern era, and vice versa.[25] The

---

21. Rasmussen also notes connections between Descartes's philosophy, the "nature as machine" metaphor, and the development of factors that facilitate the rise of modern market systems. See Rasmussen, *Earth Community, Earth Ethics*, 119. For more on Descartes's enduring legacy on modern thought, see also Rasmussen, *Earth-Honoring Faith*, 295–300.

22. Nash, *Loving Nature*, 75.

23. Indeed, the human affect on climate would not be introduced until Alexander von Humboldt's work in the mid 1800s. See Wulf, *Invention of Nature*, 213.

24. Gottlieb, *Greener Faith*, 72.

25. Johnson, "Losing and Finding Creation," 10. For more on Descartes and Christianity, see Steiner, "Descartes, Christianity, and Speciesism," 117–31. For more on Bacon and Christianity, see Wybrow, *Bible, Baconianism, and Mastery*.

cleaving of the world of nature from the field of theology, alongside the rise of overly reductionist modern scientific philosophies, contributed to a view of the world as a machine filled with mere objects holding no value beyond their utilitarian services to the goals of men.[26]

All the while, the modern era experienced a surge of diverse streams of thought emphasizing human agency. For example, the general focus on the individual's relationship to God through Christ during the Protestant Reformation and Martin Luther's (1483–1546) theology of the person as justified by faith alone stressed the individual's standing before God. While John Calvin (1509–1564) and others did maintain some emphasis on the doctrine of creation, by and large, mainline Protestant traditions in the 18th and 19th centuries came to underscore theology as the clarification of the God-human relationship. The Kantian (Immanuel Kant 1724–1804) turn to the human subject as the locus of rational thought further entrenched this stress on the human person.

With the Industrial Revolution, shifts to new forms of labor occurred along with changes in the way work was perceived and moralized by Christian traditions, especially Protestantism.[27] An unprecedented increase in the generation of wealth was a noteworthy consequence of this era. In 1786, John Wesley pondered the implications of that increasing wealth among the Methodist faithful. He feared that, with few exceptions, "wherever riches have increased [. . .] the essence of religion, the mind that was in Christ, has decreased in the same proportion."[28] He believed that religion rightly inspired a Christian commitment to personal industry and frugality leading to the accumulation of wealth, but warned that such wealth resulted in real moral dangers for the Christian, including a tendency toward "pride, anger, and love of the world in all its branches."[29]

---

26. See, for example, Descartes, "Animals are Machines," 274–78.

27. Expansion of the extractive economy and the exploitation of natural resources fed booming coal, iron, and steel industries in particular. Expansion of industries such as the textile industry saw many people shift to factories as the primary locus of work life from the home-based cottage industries of earlier periods. This, along with a growing merchant class, and increased commercialism resulting from expansion of the railroads connecting resources with consumers, demonstrates the emergence of many new forms of labor and wealth creation in the industrializing West. Population, production, and consumption levels soared. See Appleby, *Relentless Revolution*, 23–26.

28. Wesley, *Works of John Wesley*, 529–30.

29. Wesley, *Works of John Wesley*, 529–30.

For Wesley, the problem was not the industrious nature of work or the frugality inspired by a religious modesty of fiscal temperance and prudent spending. He believed leisure time could be dangerous because it risked idleness. He worried that the profligate use of one's resources could lead to a love for the world and of worldly desires that would distract Christians from the gospel. "We *must* exhort all Christians" as Wesley professed, "to gain all they can, and to save all they can—that is, in effect to grow rich!"[30] The only way Wesley imagined Christians could mitigate the problems accompanied by becoming wealthy was by urging that "[i]f those who *gain all they can*, and *save all they can*, will likewise *give all they can*, then the more they gain, the more they will grow in grace, and the more treasure they will lay up in heaven."[31] In short, Wesley was suggesting a way to avoid an early form of conspicuous consumption by redirecting excess wealth toward good works for the community.[32] This was perhaps an early effort to draw moral boundaries around the use and abuse of excess capital among the Christian laity.

Max Weber saw in Wesley's exhortations, and in the rise of Protestantism more broadly, a general affirmation of wealth as a sign of God's favor.[33] Weber argues that Wesley taught on the moral consequences of

30. Wesley, *Works of John Wesley*, 529–30.

31. Wesley, *Works of John Wesley*, 529–30.

32. Wesley, *Works of John Wesley*, 529–30. Interestingly, what may have been a problem of conspicuous consumption for Wesley, today seems to be compounded by what is now a resource-constrained planet and a quickly increasing global population. See, for example, Durning, *How Much is Enough*, 30–36. He argues that "mass consumption came of age" in the United States shortly after the end of World War II and since then, "conspicuous consumption" has spread across much of the world without regard for some of the negative consequences of the "consumer lifestyle" upon traditional cultural values of frugality and with regard to environmental harm.

33. Weber identifies a number of factors he argues contributed to the rise and success of early capitalism. First, he makes reference to Adam Smith's idea that the division of labor is an important element of advanced capitalism. Second, he stresses that the generation of wealth played a huge role. Finally, these factors were joined by the rise of Protestantism. See Weber, *Protestant Ethic*, 107. See also the first chapter of Book I of Smith, *Wealth of Nations*. Others emphasize the advancement of society with the increasing division of labor. See Durkheim, *Division of Labour*. Weber's account has been challenged and one economic historian to do so is Gregory Clark. For his alternative argument on the rise of capitalism, see Clark, *Farewell to Alms*. As Rasmussen notes, however, there are important similarities between Clark's own conclusions and Weber's. Some aspects of Clark's theory do not address important aspects included in Weber's analysis. For Rasmussen's helpful summary of divergence and convergence in Clark and Weber's arguments, see Rasmussen, *Earth-Honoring Faith*, 400–401.

industry and frugality in a way that provided a particularly religious milieu for generating the kind of excess wealth Weber argued was necessary for capitalistic systems to rise and then expand.[34] Holding labor and frugality as virtuous or morally good, Weber argues, gave rise to a broadly influential view that wealth may have been considered an outward sign of God's favor. Max Oelschlaeger writes that "worldly success, rather than being prohibited by holy sanction," became religiously reinforced during this period because "[t]he Protestant believer saw no surer indication that one was chosen (predestined for salvation) than the accumulation of wealth: economic success was a sign of God's favor."[35] The image of the self that prevails is part of a new set of stories that both powerfully shape and are shaped by the larger arc of history in the West.[36]

We see different iterations of the person in relation to God and the world accompanying the division of science and theology, the Industrial Revolution, and the rise of the Market system. One of them has particularly treacherous consequences. In it, the pursuit of wealth and its generation is enshrined as an end in itself, lending something of a religious zeal to those who proffer certain economic policies and market systems as inherently good, natural, or otherwise inevitable.[37] The rugged individual self, whose primary identity is wrapped up with the ability to overcome the fallen depravity of sinfulness through hard work and sheer determination, is one that is not adequate today even if it ever really was so. Another is the idea of the human person as a kind of *Homo administrator* or management species; a concept of the person, dependent upon the machine metaphor

---

34. Weber, *Protestant Ethic*, 118–19.

35. Oelschlaeger, *Idea of Wilderness*, 75. As per Oelschlaeger, see Troeltsch, *Protestantism and Progress*. See also Williams, *Wilderness Lost*.

36. Daniel Bell argues that before the middle of the twentieth century, the "basic American value pattern emphasized the virtue of achievement, defined as doing and making, and a man displayed his character in the quality of his work. By the 1950s, the pattern of achievement remained, but it had been redefined to emphasize status and taste. The culture was no longer concerned with how to work and achieve, but with how to spend and enjoy." In other words, Bell argues that the very wealth and prosperity that emerged from capitalist enterprise, and which is often attributed to the virtues of hard work and delayed gratification, creates so much wealth and prosperity that those virtues become undermined in the shift toward conspicuous consumerism. See Bell, *Cultural Contradictions*, 70.

37. For example, consumerism is so emblematic of free-market societies that some consider it akin to a powerful "global religion." See Loy, "Religion of the Market," 275–90. See also Porter, "Religion of Consumption."

of the natural world, which fails to acknowledge humanity's persisting dependence on and embeddedness in ecological systems.[38] Such stories about what it means to be human and how we ought to live in the world hinder us from seeing who we really are and keep us from reimagining what we could become.

The kind of religious symbolism employed by many of the world's religions, including Christian traditions, has a crucial role to play in getting us from where we are to where we could be. Religious symbolisms act powerfully and influentially upon people's emotions and motives.[39] In a broad sense, it helps people create a sense of meaning and order in what may otherwise be perceived as a meaningless and chaotic universe—it helps people make sense of their world and their place in it.[40] This is demonstrated in the way "wisdom traditions," as the world's religions are sometimes called, have often been engaged to inspire and motivate grand movements in human history.[41] Mohandas Gandhi appealed to Hinduism, the predominant religion of his beloved homeland, to push forward India's independence movement. Martin Luther King Jr. appealed to the ancient Biblical story of the Exodus to help mobilize the civil rights movement in the United

---

38. I am referencing Dean Bavington's use of this phrase, when he argues that by "framing the ecocrisis as a problem amenable to managerial solutions, deeper questions surrounding how we ought to be living and how we would like to live in the future are obscured by shallow attempts to survive in the context of the crisis-ridden ecological and socioeconomic status quo." Bavington, "*Homo administrator*," 133–4. He uses the term *Homo administrator* to denote a view of the person as responsible for fixing and optimizing malfunctions and breakdowns in "spaceship earth." See Bavington, "*Homo administrator*," 121–36, in Heyd, *Recognizing Autonomy of Nature*.

39. One of the most prominent cultural anthropologists, Clifford J. Geertz, famously described religion as "a system of symbols" that establishes "powerful, pervasive, and long-lasting moods and motivations." See Geertz, *Interpretation of Cultures*, 90.

40. Peter L. Berger's social theory of religion notes the special way in which religions act to construct a cosmos of meaning—one that he argues creates a sense of order in response to an otherwise excruciatingly chaotic universe. According to Berger, on the deepest level of meaning, chaos is the opposite of a sacred cosmos, expressed in several cosmogenic myths. He argues that religion is an "enterprise by which a sacred cosmos is established," and by sacred, he means a cosmos in which is perceived "a quality of mysterious and awesome power." By way of several examples, Berger contends that certain uniformities can be observed cross-culturally regarding the way in which that sacred quality is attributed to anything from objects and animals to people, institutions, and even cosmic forces. He posits that such a perception of the cosmos "as an immensely powerful reality" works to locate the person within "an ultimately meaningful order." See Berger, *Sacred Canopy*, 25–27.

41. Spretnak, *States of Grace*, 9.

States.⁴² When religious symbols give shape to stories of deep meaning as a response to living problems, they have the potential to powerfully shape the arc of history.

The task, then, is to rally religious ideas and stories to the challenges of a world plagued by vast disparities in human flourishing and unprecedented environmental degradation. What is going on in the world now demands a transformation of our basic understanding of who we are and how the world works. There is a gap between what is really happening in the world and the stories people tell themselves about who they are and how they ought to live. Bridging the gap between what is and what ought to be, requires first an understanding of what is really going on in the world. We need to acknowledge the breadth and depth of the problems we are confronting by looking them squarely in the face. Then, we need to decide it is not an option to run from those problems, no matter how tragic or arduous they may appear. We need bold new stories able to inspire the moral courage for us to plow ahead collectively.

Christians are called to be responsive and responsible. Religious ethics is less about prescribed moral codes and more about a critical awareness of oneself in the world, so one may more responsibly engage the world. Bernard Häring, from his Roman Catholic perspective, describes a central part of the Christian's moral life as that of learning to listen to "God's call" and then in responding accordingly.⁴³ For him, the process of becoming more ethical persons is a process of becoming more responsible persons. H. Richard Niebuhr, from his Protestant perspective, offers a view of Christian ethics as the search for what a Christian believes to be God's activity in the world, so engagement with the world comes through a deepening knowledge of oneself in relation to one's evolving knowledge of the world.⁴⁴ Both Niebuhr and Häring offer a vision of the Christian moral life as one that is attentive to whatever it is that is going on in the world. This is critical because humanity's impact on the planet has expanded so greatly that it has become a major part of what *is* going on in the world; ethical reflection

---

42. It has been argued that the Exodus story of a people's liberation has become a prototypical narrative shaping the "cultural consciousness of the West." See Walzer, *Exodus and Revolution*, 7.

43. Häring, *Law of Christ*; Häring, *God's Word*.

44. Gustafson, "Introduction," 14–16.

on humanity's influence on planetary systems simply must become a more central part of the Christian moral life.⁴⁵

While Häring and Niebuhr expand the idea of what it means for Christians to be more responsible members of society, it is also necessary to expand both the way we think about responsibility and how we think about ourselves as individuals. We need to better consider the inherently ecological nature of human embeddedness within creation. The decisions made by human beings within our social institutions are no longer only a matter of life and death for vulnerable human beings. Our decisions now affect all forms of life at the margins, struggling to survive across the planet. The present scope of human responsibility requires us to address how we might become responsible members of the larger Earth community. Moreover, as Sallie McFague argues, people in the West have come to imagine themselves as "individuals who have the right to life, liberty, and the pursuit of happiness."⁴⁶ The emphasis on individual autonomy as a primary concept of the self constitutes an impediment to more integrative concepts of the self that could better expand the frame of moral responsibility. We cannot survive and thrive independently from the societies that nurtured our humanity or the larger life support systems of the Earth that brought us into being.

We must reimagine who we are in relation to others and the Earth and in ways that are more fitting to the world, as we now understand it. Religious traditions have a rich history of shepherding powerful visions of what it means to be human in relation to others and the world. The world's wisdom traditions can continue to act as an incredible source of creativity and inspiration for the new kind of profoundly meaningful stories necessary in the transformation of human and Earth relations today. But let us be quite clear about the tremendous difficulty of this kind of transformation. The stories people use to make sense of the world do not change quickly or easily.

---

45. Jonas, *Imperative of Responsibility*, 6–7.

46. McFague, *New Climate for Theology*, 47–48. The European Enlightenment movement of the eighteenth century has had an ongoing and pervasive influence on dominant streams of Western thought. Its focus on empiricism and reductionism influenced not only the trajectories of scientific investigation and political discourse but also theology and philosophy. We might rightly celebrate some of its gifts, like a shift toward constitutional governments as well as the emergence of a robust human rights tradition. Other aspects of the Enlightenment legacy require reevaluation. It is no longer enough to envision our humanity merely in terms of individuality and autonomous rights.

The good news is that among the world's religions, including among the various expressions of Christian faith I will refer to broadly as Christian traditions, there is an incredible diversity of symbols, ideas, experiences, rituals, and stories to draw upon. Theologian Elizabeth Johnson points out that several of the world's religious traditions have in fact been creation-centered for the majority of their existence. Christian traditions, in particular, may be regarded generally as having been creation-centered for the first 1,500 years of their history and human-centered for merely the last 500 years of it—a relatively recent turn in the longer arc of Christian history.[47] Christians can rely on those historic resources by critically reclaiming them for the present and creatively reimagining them for the future—a constructive act of bridge building spanning generations of Christian intellectual heritage. It can also inspire new bridges between the insights of science and the wisdom of religious experiences, and new bridges between the world as it really is and the world as it could be. That is to say, a new world is possible—one in which Christian communities join with others so all people and all forms of life can survive and thrive together, long into the future.

A COMMITMENT TO ALL LIFE AT THE MARGINS

A concern for the poor and oppressed has to be a necessary part of the environmental movement. It is not enough to sit with the cries of the Earth without also sitting with the cries of the Earth's poorest people. Conversely, it is no longer enough for those religious traditions that have long expressed concern for the poor to do so separately from oppressed ecological systems. Christian traditions, and liberation theology especially, call the privileging of concern for the poor and oppressed within theological reflection a preferential option for the poor—it ought to go hand in hand with the concern for Earth and for all forms of life on this planet to persist and to thrive.[48] Human equity and ecological sustainability are intermeshed and need to be addressed together. Christian traditions need to do a better job of listening

---

47. Johnson, "Losing and Finding Creation," 4.

48. Liberation theology in particular has developed and insists on a need for a preferential option for the poor and oppressed in Christian ethics. See Gutiérrez, *Theology of Liberation*; Cone, *Black Theology of Liberation*; Pope John Paul II, *Encyclical Letter Centesimus Annus*, paragraph 57; Pontificium Consilium de Iustitia et Pace, *Social Doctrine*, paragraphs 182–84.

to the cries of the Earth, and the larger environmental movement needs to do a better job of listening to the cries of the poor.

Without an expressed preferential option for one set of values, history seems to show that preference is given anyway. When the implicit is not made explicit, most often the interests of those at the centers of power are protected and preserved at the expense of the vulnerable and those at the margins of society. In naming and creating space for the needs and concerns of those who are most vulnerable, like the poor and oppressed, Christian communities endeavor to occupy privileged spaces in society with Christ's radical message of liberation and inclusivity. These are spaces that people of faith rightly believe ought to include room for those historically relegated to the margins of society.

The problem, according to liberation theologian Miguel A. De La Torre, is not the expressed privileging of concern for the poor and oppressed, but in the indiscriminate privilege of some groups to prioritize and value their concerns at the expense of others. To choose, as he writes, "one ethical precept over another justifies those who will eventually benefit from what is chosen," so the goal of a preferential option for the poor and oppressed is a direct reflection of the moral commitment to love one's neighbor while embodying Christ's message of liberation and radical inclusion.[49] A preferential option for the poor and oppressed has been a central affirmation of most liberation theologians for the last half-century, and for a good reason—in an increasingly complicated and constructed world, injustice becomes more systemic and so new ways of naming and claiming justice have been needed.[50] Nonetheless, the ancient call for justice on the part of the poor is deeply rooted in a close reading of Biblical stories highlighting the sometimes radical, and always-disruptive nature of the kind of love and justice preached by Jesus and the early church in the name of those everyone else seems to forget.

A preferential option for the poor can and should extend a long-established religious concern for the world's poorest people to include other species and entire ecosystems on the verge of collapse. Is it not possible to reflect on humanity's collective disregard for the biosphere and imagine

49. De La Torre, *Doing Christian Ethics*, 12, 14.

50. See Hebblethwaite, "Liberation Theology," 179. The text references a letter by Pedro Arrupe to the Jesuits of Latin America in 1968 as the first time 'liberation theology' came into being; See also Gutiérrez, *Theology of Liberation*; Gutiérrez, "Preferential Option"; Groody and Gutiérrez, eds., *Preferential Option*; Cone, *Black Theology of Liberation*, 2–3, 120–21.

Christ's radical message of liberation might also extend to the rest of the world that Christians believe God so loves? When Christians endeavor to include the needs and concerns of the most vulnerable members of society in their theological reflection, should they not also include concern for other vulnerable forms of life on this planet? The message of Jesus carried forward by his followers in the early church is one centered on liberation for the imprisoned and inclusion for the outcast. It is a radical message in which the prevailing social hierarchies that privilege and exclude are flipped entirely on their head, and those at the margins are drawn into new, participatory roles in the life of the community.

If a preferential option for the poor is to be a practical theology for change in the world today, then Christians must acknowledge the way our humanity is bound up with all other life on this planet. Care for the Earth and care for the poor are two sides of the same coin. "[A] commitment to ecology and the environment is an expansion and undergirding of an ongoing commitment to protecting and caring for all forms of life at the margins," writes Fr. Michael J. Garanzini, SJ. This includes "the poor, the disenfranchised, the alienated and ill, the aged and disabled, new life and old life—it is an embrace of the fragility of life and a call to protect it in all its forms."[51] This is an essential expansion of liberation theology rooted in the fundamental promise embodied in the life and ministry of Christ. It invites us to reimagine the way we live in relation to others and to take a stand for justice with all life at the margins.

BOLD FAITH AND MORAL COURAGE

Christian stories about what it means to be human in relation to others and the world are sometimes able to inspire the best in people, but not always. Regarding the contemporary ecological crisis, Lynn White argued that Christianity "not only established a dualism of man and nature but also insisted that it is God's will that man exploit nature for his proper ends."[52] "Especially in its Western form," he wrote, "Christianity is the most

---

51. Fr. Michael J. Garanzini, SJ made these remarks as part of a Faculty Convocation in 2012 at Loyola University Chicago when he was then the university's president and Secretary for Higher Education for the Society of Jesus. See http://www.luc.edu/president/communications/facultyconvocation/archive/facultyconvocation2012/ (accessed September 25, 2013).

52. White Jr., "Historical Roots," 1205.

anthropocentric religion the world has seen."[53] While White's argument oversimplifies thousands of years of religious history, Christian traditions have been at least partly complicit in the creation of the contemporary ecological and climate crisis.[54] For theologians such as James Nash, this warrants a "confession of sin" that acknowledges that complicity.[55]

The lasting consequences of Christian theological and philosophical complicity in the ecological crisis amount to what has become a double problem: the degradation of the planet as well as the persisting, collective inability to take the serious action necessary to stop it. As William C. French describes it, the second part of this double problem occurs primarily because society lacks a story that is "fitting" for the unusually large scope of the problem.[56] In other words, we need to learn to think bigger about the stories that govern our lives than perhaps we have in the past. But the problem is doubly difficult in another sense as well. Leonardo Boff describes the concerns of both liberation theology and ecological discourse as stemming from "two wounds that are bleeding."[57] The first wound he calls "the wound of poverty and wretchedness" that "tears the social fabric of millions and millions of poor people the world over."[58] The second is "systemic aggression against the earth," that "destroys the equilibrium of the planet."[59] We need to tend to both wounds, and it will not do if our response is listless and faint-hearted.

The stories people tell themselves, including those Christian stories about who we are and how we ought to live and work in the world, matter profoundly. Some of them need to change. At this critical moment in human and Earth history, every idea needs to be considered for what it might contribute toward a more just and sustainable world. That includes ideas about what it means to be more fully human on a planet in which the future of all life on Earth hangs in the balance. There are billions of Christians across the globe, some of whom are grasping for a scientifically informed

---

53. White Jr., "Historical Roots," 1205.

54. Nash, *Loving Nature*, 72–74.

55. Nash, *Loving Nature*, 72–74.

56. French references Aristotle's argument that in ethics it is necessary to attend to "ultimate particulars" that are "fitting" to a given context. See French, "On Knowing Oneself," 162, 165. See also Aristotle, *Nicomachean Ethics*, 157, 160.

57. Boff, "Liberation Theology and Ecology," 134–39.

58. Boff, "Liberation Theology and Ecology," 134–39.

59. Boff, "Liberation Theology and Ecology," 134–39.

spiritual story that can help them make better sense of their faith, the world as it is, and the many problems before us than do the stories they may have inherited from their religious traditions. They may already see some inherent problems and contradictions in those stories. They long to really see things as they are and not merely as they have been taught to perceive them to be. If they engage in serious, critical theological reflection in light of the ecological and climate crisis before us, there could be a flood of new stories of the kind the world so desperately needs. That cannot happen, however, unless people of faith are willing to embrace this great work audaciously.

It is hard to be bold; it is hard to both work within traditions and challenge them at the same time. But, it may be the most potent way forward, and incredibly fruitful if we are brave enough to attempt it. As a gardener, I approach much of the world through a gardener's lens. I cannot help but attribute certain qualities to the plants I tend. I have grown peonies for years, and they are one of my favorite cultivated flowers. I cannot help but marvel at the peppy optimism of their buds almost as much as I enjoy their crass, overly enthusiastic, wildly explosive blooms. When they bloom, they seem to be shouting to the world that this is their moment and they are going to give it all they've got! Never mind that the first spring rain is probably going to obliterate them. It is as though they are stepping into the unknown with a bold faith in which they have nothing to lose for trying. That audacity of theirs makes the world a richer and infinitely more beautiful place, even if only for a moment. I am bewitched by their "I don't give a damn" "this is do or die" attitude. I think this could be a model.

I wish more Christians could approach their faith with this kind of audacity. I wish more of us could approach the world with the faith of a peony—life would surely be a more magnificent ride and the world would probably be a more beautiful place as well. Too often we humans approach the uncertainties before us with an attitude of fear and hesitation. We do not go boldly into the night.[60] We do not approach the mysteries of life with the sense of awe and curiosity rightly proportionate to them. We are too often distracted by the mundane banality of everyday life. Instead of seeing those events and interactions as part and source of those larger mysteries that shape the cosmos and our lives within it, we allow them to distract us from what really matters. That is, we do not tend to fully engage the world as it is but rather the world as we can bear it. We gravitate toward thoughts

---

60. This is a reference to a poem by Dylan Thomas called "Do not go gentle into that good night," in *Botteghe Oscure* (1951), 208–10.

we can understand, positions we can control, ideas we can wrap our arms around and then comfortably place in a nice little box so we can move on to other things. The stories we have inherited help us do just that. New stories unsettle us, and we need to be unsettled.

The lens through which I see the world rests in one fundamental assumption: if there is a God, then God delights in this good Earth, and any sense of God's presence is diminished in its degradation. Human poverty is an especially cruel injustice, and the poor tend to suffer disproportionately from environmental degradation. Implicit in my understanding of whom God might be is a call to radical inclusivity for all who are marginalized and excluded by unjust systems of poverty and oppression. A preferential option for the poor and oppressed, as Jorge V. Pixley and Clodovis Boff so eloquently write, is ultimately about how one orients oneself in response to a belief in a God of love and justice who is the first to opt for the poor.[61] If this is indeed the case, then let whoever so believes boldly march directly into the pain and suffering of this world knowing they bear the heart of the Universe in their being.

---

61. Pixley and Boff, *Bible, Church, and Poor*, 109.

2

Facing the World as It Is

AT A SMALL, MATERNAL health clinic in rural Bangladesh, a brief encounter with a little girl named Hawa transformed the way I see the world. When Hawa was admitted to the clinic, she was one month old and immediately diagnosed with kidney failure as a result of severe malnourishment, undernourishment, and dehydration. Her mother, who was jaundiced and visibly emaciated as well, was unable to make the breast milk her baby needed to survive. Like many other women in the village, Hawa's mother did her best to care for her daughter with the few resources she had at her disposal by feeding Hawa a watery mix of sugar and rice powder. When the milk substitute proved inadequate, Hawa's family made the trek to a clinic to seek medical care they knew they could not afford. The medical staff at the clinic were able to save baby Hawa's life by admitting her and her mother into the hospital without a request for payment. While Hawa is one of the little girls who will live to fight and, I hope, win many more battles over the course of her life, the tragic reality is that so many other women and girls have not had and will not have that opportunity.

So many women and girls lack access to basic but necessary medical care as a result of social inequality that in places like Bangladesh "missing women" number in the millions.[1] Insufficient and unequal healthcare is partially responsible, but so too are other factors related to a general lack of women's agency. In those places that value men and boys more than girls, generally boys are fed first, educated first, and receive care first. To the ex-

---

1. Sen, "Missing Women," 587.

tent women and girls are not accorded any agency in decisions regarding their bodies, the men in their lives often decide whether they live or die.

The image of Hawa and her mother clinging to life in that small clinic, seemingly forgotten and unnoticed by the rest of the world, sticks with me. I can no longer say I see the world as simply a beautiful place where everything happens for a reason as part of some divine plan. But, I cannot forget those other moments I had in Bangladesh when I paused to admire the warm hues of the sun setting over rice fields reaching toward a vast horizon. Even amidst the unimaginable despair we encounter in life, people are so often capable of finding solace, joy, and beauty, in things like a cup of hot tea shared with a friend, that first breath of crisp morning air, the sound of children playing or of the wind rustling through the leaves on a warm summer day. How do people of faith make sense of a world that appears to be mixed up with so much darkness, but also light?

There are the mountaintop experiences that offer a kind of magical encounter with the world and nurture a sense of connection to what many people might describe as a comforting, numinous presence—or for some, simply, God. If we are lucky enough to have them, they are the precious moments when the seemingly large problems of daily life melt away, if only briefly, into smallness that seems more befitting of them in that moment. For a lot of people, myself included, these encounters are God-moments; they reveal a glimpse of something utterly beyond words. Like Moses before the burning bush, we know the ground upon which we stand is holy when we are drawn out of ourselves by such experiences and into a grander vastness of being we share with others.

Still, it is a tragedy that so many little baby girls like Hawa may never live to experience this other side of life. Their only experience of the world will be short and filled with suffering. Where is God in the context of their experiences? Where are we, those of us who might be able to do something but are paralyzed into inaction amidst such an unbearable darkness? The sacramental tradition of anointing the sick with holy oil is an ancient one, rooted in the idea that God's light can be carried boldly and bodily, directly into the mystery of suffering, and persist there in the midst of that suffering. God is there even when we are helpless; it is not an excuse for inaction but an invitation to join God in co-mission, to accompany our neighbors in need so they are at least not alone in their suffering. When they cry out, we cry. When they mourn, we mourn too. When they hurt, we are hurt. God does so too, and the challenge is always to do something about it.

This means that for Christians the incarnation is not simply a historical moment in time that they believe happened some two thousand years ago. It is an ever-present aspect of everyday life. To live as a people of Easter faith is to believe that Christ is always breaking into the here and now, to dwell in and amidst the world—everywhere and in all things—calling us to meet him there at the cross in the face of our neighbor's suffering, and joy. It is meeting God where God dwells, there with us in and through all forms of poverty, all kinds of hunger and longing. God is there in and amidst the turmoil, violence, and death. God is there too in times of conflict, warfare, and all varieties of destruction. That is not to valorize such things because some people believe they encounter God in them. It is instead to acknowledge the way many conceive of God's light as a piercing interruption of an otherwise impenetrable expanse of darkness.

I hope that if it is possible to imagine a God who might love and cherish the life of a little girl like Hawa even when the rest of the world fails to notice her, we might also imagine a God who loves and cherishes all forms of life at the margins. Does God dwell in and amidst various forms of planetary suffering, inviting us to meet him there too? Species extinction. Glacier melting. Coral bleaching. Clear-cutting. As I perceive of God, I cannot help but image one who loves the world so much that God would desire to be there too. Though the church has no ritual to help visualize this mystery of faith, I wonder if this is partly what it means for Christians to step into the fullness of God's mystery during this particular moment in human and planetary history.

If Christians believe God is there in and amidst everything, then that means it is a theological activity to study the world as it is. The hardest part in facing the world as it is, is that it can feel too overwhelming to do so without becoming desensitized or crushed by the weight of the world's despair.[2] I have seen so many students paralyzed by a sense of how much work needs to be done to confront systemic injustice. I have seen the flame of passion die out in too many activists who so desperately wish to make the world a better place. At times the world as it is bears so much ugliness that we lose sight of its beauty, its magnificence. This should not stop us. Alongside the complete denigration of biodiversity and ecological systems across the planet, communities of people around the globe are suffering

---

2. I speak to this from personal experience in observing both students and colleagues who begin a study of the issues contained in this chapter, contributions of which come partly from revisions to Mastaler, "Case Study on Climate."

from all forms of extreme poverty and social inequality. This is the world into which every new generation is born. To study such things as they are is not easy, but so very important.

Before we dive into the difficult task of engaging some of these issues in the subsequent sections of this chapter, I need to introduce a few key terms. While a phenomenon such as global climate change is by its very definition a shared global problem, it is by no means a burden that all people on this planet share in the same way. A population's *sensitivity* may include those factors that weaken the standing of any particular group of people within the larger society, such as being a woman in gender-biased societies, or being poor where great economic disparities and the most extreme forms of poverty exist pervasively. *Exposure risks* include those environmental factors that directly threaten a population's ability to survive and thrive, such as natural disasters and the kind of natural resource depletion that happens when fresh water sources dry up, fisheries collapse, and fertile farmland erodes away.

Resiliency is a population's ability to mitigate challenges by reducing sensitivities and minimizing exposure risks. So, climate resiliency may be regarded as the ability to survive and thrive in the face of the many challenges that accompany climate change, such as rising sea levels, shifting weather patterns, and changes in patterns of disease transmission. *Vulnerability* is the reverse of this. Populations with ecosystem-dependent livelihoods, such as fishing or farming, are particularly vulnerable because in many places they tend to be women and poor (they have a high climate sensitivity) and they are dangerously exposed to risks from environmental changes. Around the world, people with ecosystem-dependent livelihoods generally tend to have some of the lowest-paid occupations in society. They are more reliant upon stable weather patterns, such as adequate rainfall for crop production, or particular hot and cold cycles that influence agricultural pests and crop fertility.

Climate vulnerabilities exist for sensitive populations even in more developed countries where exposure risks may be lower, and they impact the most vulnerable in those societies too. For example, the US has a sizable population employed in non-ecosystem-dependent livelihoods. Since these livelihoods are generally less dependent on environmental fluctuations, their sensitivities and their exposure risks would be considered significantly lower. Middle- and working-class families in places like the US generally have access to social support systems generally unavailable around much of the world, even though that safety net faces a near-constant onslaught of erosion from vagaries in the political process and inequitable distribution among all communities.[3] When Superstorm Sandy hit the Eastern US shoreline, an area generally regarded as relatively affluent in the US, the impact on homeowners was quite significant even despite the region's general level of wealth.[4]

The resilience of the US as a whole can act as a safety net for even large regions of the country, because of the relative size and wealth of the nation's highly diversified economy and citizenry, its large geographical territory, and influential presence in the global economy. These advantages can make institutional, infrastructural, and social services available to individuals and communities suffering from localized catastrophes, even if it does not always happen equitably after a disaster.[5] After Sandy, a massive inflow of donated resources and volunteer labor from the rest of the country also contributed to clearing and rebuilding efforts, even though the personal hardship experienced by scores of US citizens in the face of this disaster was nonetheless quite devastating.[6] When Hurricanes Harvey, Irma, and Maria hit the United States and its territories, the outpouring of resources from around the country to Texas and Florida was quick and significant (although, unfortunately, less so for Puerto Rico). Every resource matters when it comes to increasing resiliency, and it helps impacted regions to be able to draw on a larger national safety net.

Planetary denigration and systemic injustice are challenges that are connected in the way that they do not affect everyone equally. The consequences of ecological degradation and climate change are borne most heavily by the already poorest and most vulnerable on this planet, and that is because social structures and institutions do not yet adequately protect

---

3. For a general class analysis in the US, see Durning, *How Much is Enough?*

4. New Jersey Department of Community Affairs, *Disaster Recovery: Action Plan*, 2–11. Data show that "59,971 owners' primary residences sustained some amount of physical damage. Of this number, 40,466 homes sustained severe or major damage." New Jersey Department of Community Affairs, *Disaster Recovery: Action Plan*, 4.

5. Hurricane Katrina in 2005 is often held up as an example of the way poverty, race, and political processes can help some and leave others in need.

6. In some of those communities damaged by Superstorm Sandy, up to 80 percent of the housing stock was purportedly composed of second homes owned by middle-class citizens for whom their beach house was either inherited or considered their retirement savings. While allegedly inadequate, significant government subsidies and subsidized insurance monies helped to soften the economic impact of the storm on homeowners and communities. See Paik, "Left Out."

and empower enough of the world's people. This is what makes climate change both a moral issue and a social justice issue. It emerges from and must be addressed through the institutions, organizations, and structures of an increasingly globalized world. To understand an issue as complex as this one, is central to responsible theological reflection on it.

GLOBAL POVERTY AND ECONOMIC INEQUALITY

To study the world as it is today necessitates a study of global poverty and economic inequality. Almost everyone seems to have some sense of poverty's pervasiveness—we almost all certainly think we know it when we see it, and some of us may even know of it through our own direct encounters with it. Can we define it, however, in a way that helps us to see it better, or more clearly in our larger societies? Can we measure it, so we can know whether or if any of society's collective efforts to counter it are working? These are important questions because people from varying socioeconomic and national backgrounds quite reasonably define poverty differently from each other. It is much simpler to describe poverty across US cities than it is to compare poverty in the US to poverty elsewhere. Moreover, discussing poverty in neighborhoods within Chicago is more straightforward, for example, than comparing Chicago's poverty to London's. Although poverty exists in both places, they have different cultural and socioeconomic histories. Poverty is largely context-specific, and context matters.

Comparing Chicago's poorest neighborhoods to the slums of Kibera or Mathare in Kenya is more difficult still. Having grown up in the United States as what some there may consider poor, and having been witness to various forms of poverty in places across Africa and South Asia, I think some distinctions are helpful. Jeffrey D. Sachs, an economist, and director of the Earth Institute at Columbia University, describes three different categories of poverty: extreme poverty, moderate poverty, and relative poverty.[7] *Extreme* poverty is the kind in which chronic hunger, illness from unhygienic living conditions, and a lack of adequate shelter threaten basic survival. Sachs regards *moderate* poverty as a form in which daily basic survival may be assured, but it is precarious and fragile. It is only marginally better than extreme poverty. Perhaps one has access to a daily meal, but only the most basic staples; or perhaps one's lifestyle is dependent upon daily labor—if you cannot work for the day then you cannot eat that day.

7. Sachs, *End of Poverty*, 20.

Relative poverty is the form most ubiquitous across the United States and Western Europe. It is a form of poverty in which a person's ability to thrive and flourish within society is at stake, however that may be perceived within the larger society. To experience relative poverty may be to live with one's most basic needs met, one day, week, or month at a time, partly because a basic social safety net often ensures one has access to the resources they require in order to avoid extreme poverty. As Sachs points out, even though basic survival may be assured, those in relative poverty often lack access to important cultural goods—things like an equitable education and a realistic opportunity for achieving a life and livelihood that allows them to escape the near constant fear of never having enough.[8] Violence, abuse, and neglect, either in the home or in the community are far more common among the relative poor than those who can afford to move into more affluent neighborhoods. People in all three categories, although they may differ in their experience with various forms of poverty, are still poor. Poverty is everywhere, even in some of the richest cities on Earth.

Sach's categories are helpful to the extent they offer a general understanding of poverty's global pervasiveness, but they are not intended to offer a snapshot of metrics demonstrating over time what activities can be relied upon to increase or reduce global poverty. One conventionally used measure describes economic poverty in terms of an income at or below $1 a day. This has long been the most common way to measure poverty, and according to this metric the global poor live primarily in South Asia and Africa. Poverty, however, is much more complex than this and it needs to be described more concretely. What is at stake in any description of poverty is the way it partly determines who is helped and by what means. Does someone rise above their poverty if their income rises above $1 a day, even if they achieve no real gains in access to those things a higher income is supposed to buy? No, of course not.

When a Zimbabwean woman was asked how she might define poverty, she replied with penetrating clarity:

> How can you ask that question when you yourself can see that I live in poverty? The definition of poverty is already in front of you. Look at me, I stay alone, I do not have enough food, I have no decent clothes or accommodation, I have no clean water to drink nearby. Look at my swollen leg. I cannot get to the clinic as it is far for me to walk. So, what kind of definition of poverty do you

---

8. Sachs, *End of Poverty*, 20.

expect me to give you, which is better than what you have seen with your naked eyes?[9]

This woman offers a description of her poverty grounded in her direct experience with it. If it is a theological activity to study the world as it is, then her words are a more apt description of the human experience than abstract markers, which may be completely divorced from those factors pertaining to an individual's quality of life. From the Zimbabwean woman's words, we are reminded it is important to include those things an income is intended to procure: things like medical care, clean drinking water, and food.

Efforts have been made to begin describing and measuring poverty around the globe in a way that is reflective of its multifaceted complexity. The United Nations has developed the Human Poverty Index (HPI-1). It is an effort to create a measurement of poverty that assesses a person's opportunity to live a long healthy life, to pursue basic education, and to obtain the kind of decent standards of living that more developed countries have achieved, although in varying degrees.[10] It does this by including a person's probability at birth of dying before age forty. It measures adult literacy rates as an indicator of basic knowledge. It also factors in the percentage of a population not using improved water sources and the percentage of children who are underweight for their age. The index presumes each of these statistics reflects something of a country's potential to offer a decent standard of living for its citizens.[11] When the HPI-1 is mapped, it shows that South Asia and much of Africa still represent the highest concentrations of the most extreme forms of poverty. While this index only includes rather simple measurements, its strength is that it relies on empirical information that is widely available across many different countries.

A more promising index with the potential to supplant the United Nation's HPI-1, the Multidimensional Poverty Index (MPI), may one day be regarded as the premier multifactor poverty index.[12] This is because the MPI takes into account ten indicators, which is several more than the

---

9. World Health Organization, "WHO Position Paper," 10.

10. While the UN has no established convention for the terms "developed" and "developing" countries, the terms and their meaning here are broadly used to distinguish between generally more affluent countries such as the US and those in Western Europe and poorer countries such as many across Africa and some Asian countries like Bangladesh. See United Nations Statistic Division, "Composition of Macro."

11. UNDP, "Fighting Climate Change: Human Solidarity in a Divided World," in *Human Development Report 2007/2008*, 354.

12. UNDP, *Human Development Report 2010*.

handful considered by the HPI-1. An ongoing challenge is that published data on many of those indicators is lacking for a number of nations. So while the measurement helps us paint a clearer picture of poverty wherever the data does exist, the lack of complete data sets means some places are not yet measured. Still, some of what can be learned about poverty according to the MPI, is that more than half (51 percent) of the world's poor live in South Asia and over a quarter (28 percent) live somewhere on the African continent.[13]

If any of these three measurements were all we had to assess global poverty, we might have good reason to doubt their ability to tell us much of anything at all. Taken together, they lead to a stunning insight: no matter how you measure it, the most extreme forms of poverty are clearly concentrated overwhelmingly and disproportionately across the African continent, and in South Asian countries like India and Bangladesh. The concentration of such large numbers of the world's poorest people in these locations make these highly sensitive populations that are in turn more vulnerable to risk, and especially the exposure risks of environmental and climate change. The world's poorest people are disproportionately impacted by the denigration of planetary systems, and from what we do in places like the United States and Western Europe. When we peel away the veil of ignorance, we see a world we have created where so many people suffer in poverty and have so few opportunities to do anything at all about it. They have been left behind by the failures of an unjust system. It can be painful to see this, but the world must not turn away.

GENDER DISPARITIES AND SECONDARY POVERTY

Even within the most vulnerable groups, some are yet more vulnerable than others. Women, as a gender group, and especially the poorest women in the poorest, least equitable societies, are a particularly sensitive population. Their unequal social standing makes them disproportionately vulnerable to environmental and climate changes.[14] They are vulnerable because of their high climate sensitivity and their higher exposure risks. The women who

---

13. Alkire and Santos, "Multidimensional Poverty Index," 4.

14. For more on gender and social context, see Butler, *Gender Trouble*; Butler, *Undoing Gender*. See also Cahill, *Gender & Christian Ethics*; Jung and Vigen, eds., *God, Science, Sex, Gender*.

live in the world's most economically impoverished communities occupy some of the most marginalized social spaces on the planet.

While poverty reduces the likelihood of survival for both women and men during a natural disaster, being an impoverished woman makes death more likely. This devastating but simple fact plays out all across the world time and again during all sorts of disasters. Whether it is a cyclone in Bangladesh, a heat wave in Europe, or an Asian tsunami, women and girls die in higher numbers and at earlier ages than do men during natural disasters.[15] In some places, women have been up to five times more likely to die than are their male counterparts.[16]

Why is it that a person's gender is one of the most important factors in determining whether she will survive when disaster strikes? There are three factors driving the complex socioeconomic challenges confronted by women in the most unequal societies. The first is that women generally work less often for pay and receive less pay for comparable work, especially in activities like subsistence farming and water collection, which makes them more likely than men to be affected by factors like soil erosion, drought, and desertification.[17] Women also experience a type of secondary poverty, which results from being married to a man who may spend too much of a family's income on items such as alcohol, drugs, and gambling.[18] And women are more likely than men to support and head single-parent households.[19] Families face extreme hardship when they are supported by women forced to eke out a living in societies in which women's labor does not earn pay equivalent to men's or in which many or most jobs are simply

---

15. A United Nations Population Fund (UNFPA) report finds that the "case studies associated with a devastating 1991 cyclone in Bangladesh, the 2003 European heat wave, and the 2004 Asian tsunami [. . .] affirm the greater vulnerability of women" when considered along with sampling data from 141 countries between 1981 and 2002, all of which show that women die in higher numbers and at earlier ages than men in natural disasters. See UNFPA, "Facing a Changing World," 45.

16. The 1991 cyclone in Bangladesh killed five times more women than men. The study shows that the "[l]ow socio-economic status of women correlates with larger differences in death rates" such that "the more severe the disaster and the lower the socio-economic status of the population affected, the greater the gap between women's and men's death rates in such disasters as cyclones, earthquakes and tsunamis." UNFPA, "Facing a Changing World," 35, 45.

17. UNFPA, *Facing a Changing World*, 35.

18. Secondary poverty is a phrase coined by B. Seebohm Rowntree, *Poverty*. See also Townsend, "Measuring Poverty," 130–137; Veit-Wilson, "Paradigms of Poverty," 69–99.

19. UNFPA, *Facing a Changing World*, 45.

unavailable to women.[20] The simple fact is that around much of the world, female-headed households are poor, especially in the world's poorest countries where it is women who carry the daily burden of keeping their families afloat.[21]

Poverty can be exacerbated by social context, however, especially regarding the way boys and girls are raised. Certainly, medical conditions that make a person physically vulnerable, such as pregnancy, increase a person's overall vulnerability amidst disaster. Physiological factors matter; good upper body strength, regardless of sex, helps a person cling to a solid structure or climb out of danger when floodwaters rage. But, in places like rural Bangladesh where women are more likely engaged in domestic responsibilities inside the home, women are generally more likely to be trapped inside their homes when floodwaters hit. Girls are not generally taught to swim, nor do they learn to climb trees like their brothers, and so they are that much more unlikely to escape drowning amidst a devastating flood.[22]

There are also larger global forces at work in ways that are destructive to families and especially to women. Those forces increase the odds that women and girls will not survive environmental and climate disasters. Women, as well as children and the elderly, "are more likely to stay behind [in disaster-prone areas], while younger male members [up to 90 percent in some areas] are more likely to leave home."[23] This is part of a global trend that includes the push and pull factors associated with the mass movement of people into urban megacities.[24] They are driven out of their rural setting by lack of economic opportunity and they are pulled toward the fringes of growing mega-cities with dreams and visions of urban prosperity. Many young men leave their family villages behind and their wives as heads of household to look for work in larger cities, like Bangladesh's capital city of

---

20. See Lin, *Gender, Modernity*; Rajan, ed. *India Migration Report 2011*. See also Jetley, "Impact of Male Migration," WS47–WS53; Agesa and Agesa, "Gender Differences," 36–58.

21. As an example, one report references findings by the Asian Development Bank to show that, "over 95 percent of these female-headed households are below the poverty line." See Cannon, "Gender and Climate Hazards," 48. For more on sea-level rise in Bangladesh, see Huq et al., "Sea-Level Rise and Bangladesh" 44–53.

22. See Friedman, *Hot, Flat, and Crowded*. See also Kristof and WuDunn, *Half the Sky*.

23. UNFPA, *Facing a Changing World*, 35.

24. Saunders, *Arrival City*; Klare, *Rising Powers, Shrinking Planet*; Sachs, *End of Poverty*.

Dhaka. Many of these men who leave may not find the work they seek and are unable to send money home as expected, so women are left with the sole responsibility of feeding and caring for the daily needs of their family. Even when money is sent home, such urban-to-rural remittances do not always contribute significantly to rural economic development.[25]

The economic currents at play in an increasingly globalized world are such that local rural farmers can rarely compete with prices set by international agribusinesses. The offshoring of manufacturing from developed countries to urban centers in developing countries creates the appearance of economic opportunities for displaced rural workers. It also creates huge profits for international corporations who benefit from cheap labor, at least until the economic conditions evolve and necessitate future moves to even cheaper labor markets elsewhere. Such forces greatly influence mass population movements as well as dynamics between men and women. They create undercurrents that deeply complicate otherwise personal family decisions and that tends to leave women especially vulnerable in their wake.[26]

I saw many of these dynamics play out in the lives of my friends and their families during the several months I lived in rural Bangladesh while working for a local nongovernmental organization devoted to women and children's health care, community development, and disaster response programs.[27] I spent most days teaching English to the staff so they could work with visiting physicians from the US and around the world. This experience also allowed me to observe firsthand the challenges of community-based

---

25. See Rempel and Lobdell, "Role of Urban-to-rural Remittances," 324–41. This is, however, not always the case and in some places, like Bangladesh, remittances from abroad can reach sums greater than the country's entire export earnings or all foreign aid coming into the country. See Saunders, *Arrival City*, 29–30.

26. See Friedman, *Hot, Flat, and Crowded*. See also Kristof and WuDunn, *Half the Sky*.

27. In a position funded by the Evangelical Lutheran Church in American (ELCA), I worked for Lutheran Health Care Bangladesh (LHCB), which is an international development and humanitarian organization that describes itself as operating "health, development and disaster response programs in different districts, especially in disaster prone areas of Bangladesh. [. . .] LHCB operates an 'Integrated Health and Community Development Project' through its hospital at Dumki, in the Patuakhali District," and it emphasizes "the concept of Integrated Community Development [through] various income generation activities, capacity building and women empowerment programs, formation of people's organizations along with water and sanitation activities to ensure holistic development for both the individual and community as a whole." See "LHCB in Brief" at http://www.lhcb.org.bd/?p=lhcb_in_brief (accessed 5 July 2016).

capacity building programs as well as women's health and economic development programs in a rural, third world context.

I spent most evenings socializing with the other young men in my village at the local tea stand. Many of those who had not yet left the village for the capital city planned to do so at some point in the future. Several of these young men subsisted on one large meal a day, consisting usually of white rice, bananas, and several quarts of water to help the belly feel full (or so I was told). Those young men were quite forgiving of my limited Bangla language abilities. I had no Internet connection, no email, no television, so they became my primary source of social interaction beyond the Bangladeshi hospital staff and its patients. I do not remember women typically venturing out after dark and when they had to, they did not generally stop to talk with us. So these young men and I talked a lot. I will be forever thankful they trusted me enough to invite me into their lives and allow me to learn from them and their stories, their worries, and their dreams.

The memories of the people I met in Bangladesh stick with me, and they are at the front of mind every time I review the kind of statistics around poverty and gender disparities I have asked you to consider in this section. Real people suffer from the global and social dynamics our societies have constructed. It is the unyielding sense of hope I saw in those who suffer most from these dynamics, however, which continues to stand out even more clearly in my memory. It reminds me that studying the world as it is, as a theological activity, is also about seeing hope persist in spite of the most painful circumstances. It is to seek out a light in the darkness and resist the forces that might otherwise snuff it out. To engage in Christian ethics is to engage the world, as it is—it is to acknowledge its suffering, and to respond with bold faith and moral courage.

SOCIAL AND ECOLOGICAL CONCERNS ARE CONNECTED

Many of the most pressing social injustices are as rooted in ecological problems as are ecological problems rooted in global systems of injustice. Rarely can issues arising in human social systems, such as food insecurity and human health inequalities, be treated adequately without addressing issues related to Earth's ecological systems. To resist the structures of injustice and oppression that marginalize the poorest and most vulnerable members of society, is to resist structures that degrade and deplete the ability of all forms of life on this planet to survive and thrive into the future. If a case is to be made for an ethic of responsibility that attends to human welfare, then the case for environmental and ecological concern needs to be made too. The two are inextricably linked.

The contemporary context requires a collective level of cooperation in our care for and attention to the Earth and its ecosystems, which has never before been attempted by our species. The general trajectory of humanity's relations to other species is one in which most other species often must yield their very existence to human activity. The current extinction rate has been exceeded only a handful of times in the totality of Earth's history. Five great extinction events in Earth's history can be attributed to rapid, large-scale geological and atmospheric changes such as meteorite clusters, asteroid impacts, and major volcanic events. What is different about then and now is that for the first time in Earth's history, a single species has become a force akin to whatever it was that killed the dinosaurs.[28] To significantly alter our current trajectory will not be easy. It very well may take several generations, if it changes at all. But, human welfare is ultimately tied inseparably to planetary well-being. Real conflicts do exist between impoverished communities, other species, and ecosystems, namely through the pursuit of resources necessary to develop and thrive in ways that may be antagonistic to the continued functioning of intact ecosystems. While the goals of economic development and conservation may seem to be mutually exclusive at times, sustainable development sees them as linked. There are worse and better ways for societies to grow and harness their technological power. The best way forward prioritizes short-term needs within the larger context of longer-term sustainability.

All life on Earth is increasingly at the mercy of decisions made by our species. This does not change the simple fact that if we destroy our planet's ability to sustain and nurture life, we too perish. Humanity's ecological embeddedness is as real as it has ever been, even if trends toward urbanization and reliance on technology place many of us at a comfortable distance from the natural processes of our environment. The social systems we depend upon for that comfort are increasingly becoming arbiters of justice for the rest of life on Earth.

This means that when governments weigh the critical need for electrical power generation in places like Bangladesh, for example, and choose

---

28. Wilson, *Diversity of Life*, 32, 191. See also Ehrlich and Ehrlich, *Extinction*; Leakey and Lewin, *Sixth Extinction*; Ward, *Under a Green Sky*; Brown, *Plan B 4.0*; Barnosky et al., "Earth's Sixth Mass Extinction," 51–57.

whether to build a coal-fired power plant in the Sundarbans—a delicate primary forest and mangrove ecosystem—they determine whether the endangered Bengal tiger loses the critical habitat its species requires for continued survival.[29] Who am I to say, writing beneath artificial lights on an expensive computer in my comfortably heated and cooled home in a locale powered mostly by coal and nuclear energy, that Bangladeshis should be denied access to the basic electricity a plant like this would offer? All the while my heart aches at the prospect that such a fragile ecosystem will be degraded, hastening the demise of such an iconic species.

If we first acknowledge the indivisible necessity of both economic development and ecological conservation for shared planetary flourishing, then we have our twin pillars of sustainable development. Moreover, this kind of prioritizing needs to extend across the spectrum of humanity's collective decision-making to include the way the world's governments approach how land and resources are used or preserved. Governments can use this framework as one way to hold industries and corporations accountable for the larger common good. Also, it applies to universities and their curricula, campuses, and endowments; to international lending institutions like the World Bank and the projects they fund; to judiciaries and elected bodies, their constitutions and governing bodies. None can be exempt from the demands of social justice and the call to ecological responsibility because human civilizations have already become arbiters of justice for countless communities and forms of life across the planet.[30]

As arbiters of justice, human societies and social institutions must be held accountable for their continued disregard of vulnerable and sensitive populations, other species, and ecosystems. The world must ask whether our representative governing bodies act as conduits of justice or as conduits for systemic oppression, subjugation, and domination.[31] There are streams of injustice connecting the systemic exploitation of all the world's most vulnerable populations—human and otherwise. It is our collective choice whether we will continue to pretend there is no problem or own up to it and look it squarely in the eyes. If we can confront these painful realities, find

---

29. Degradation of the Sundarbans ecosystems could occur through development projects like this, but also through deforestation, increased air and water pollution, and barge traffic, all activities in the region that might, at least temporarily, increase economic activity but which directly conflicts with the need to conserve these ecosystems and the various species they support. See Hance, "Key Mangrove Forest."

30. Berry, *Great Work*, 4.

31. Nickel and Viola, "Integrating Environmentalism," 265–73.

hope in the face of them, and seek out a new way forward by disrupting the pathways of injustice, then we are one step closer to building what Christians call the peaceable kingdom of God, right here on Earth.

Such a world is only possible through the transformation of social structures and institutions, especially of those prevailing economic institutions and corporate enterprises that have come to decide who wins and loses, in a system that privileges profit for so few at the expense of so many. To create a world increasingly governed by a vision of justice for all is surely an enormous challenge with much standing in our way, but the barricades are not insurmountable. Economic oppression, gender inequality, and ecological domination are deeply embedded in social systems across cultures. But, we created these things, and so I believe—and I will remain relentlessly hopeful in this regard—that we can tear them down and build a better world.

CLIMATE CHANGE IS GLOBAL AND LOCAL

Climate change manifests not only in terms of global consequences but also as local impacts for many of the world's poorest and most vulnerable regions. We know that some of the most extreme forms of poverty exist in particular places across the globe, in places like continental Africa and South Asia. Moreover, the regional and continental climates in these places are already experiencing observable changes that are exacerbating poverty. Droughts and desertification that plague many places in Africa will become more common and more severe across much of the continent due to rising temperatures and depleted water sources. Flooding and waterborne diseases are expected to increase across much of South Asia, due to climate change-induced factors like the increased melting of snow and ice in the Himalayas. These problems are expected to only get worse for every degree of planetary warming.[32]

Conversely, climate change-inducing carbon emissions are locally emitted but globally consequential. This is important because carbon emissions are different from what may be considered more traditional types of pollution. When manufacturing processes at a factory create toxic chemical by-products that the factory then discharges into a local pond or river tributary, that kind of pollution is known as point-source pollution. Point-source pollution almost always tends to cause environmental problems at

---

32. The World Bank, *Turn Down the Heat*.

the immediate point of discharge or relatively close to the original point of discharge and the downstream or downwind area.[33] While carbon emissions act similarly to traditional pollutants in that they may be discharged from a single point, such as a smokestack or tail pipe, they have a global accumulative effect. That means that emissions on one side of the world can affect communities on the other that may release a low amount of carbon pollution.

The speed and intensity of human-induced climate change that is now occurring will almost certainly exacerbate many of the inequalities between nations that act as a point of conflict among international negotiations.[34] This is why so many argue that the question of climate culpability—who is to blame for the overwhelming bulk of climate emissions—ought to be more prominent in determining how the global community responds to climate change in terms of mitigation and necessary reparations. For some of the most resilient populations, or those with the greatest access to natural, social, and financial resources, there are more options about whether to abandon the most devastated, disaster-prone areas or attempt recovery. Even though those same nations are often most responsible for the bulk of historical emissions and may experience some negative effects in the near term, they have the best chances of surviving and thriving under a new climate change paradigm. Meanwhile, those who are most vulnerable and least responsible for causing climate change will be least able to avoid or adapt to its worst effects.

The Human Development Report Office of the United Nations describes the dramatic disparities between those who have contributed most to climate change and those expected to bear the greatest burdens. It notes

---

33. "Point-source pollution" is an industry- and government-defined term used to describe pollution discharged from a single, identifiable point. While many climate-causing emissions (such as carbon dioxide) originate as point-source pollutants and tend to accompany other pollutants (such as ash and soot) with immediate effects on the local environment and its inhabitants, carbon dioxide has a particularly global and far-flung impact when emitted in the quantities observed since the Industrial Revolution began. For several example definitions, see US Environmental Protection Agency, "Total Maximum Daily Loads"; US Environmental Protection Agency, "Waste and Cleanup"; and US Geological Survey, "Water Science."

34. There is widespread consensus among the international scientific community that environmental and climate changes are indeed occurring and that it is reasonable to interpret from the data a solid, demonstrable link between those observed changes and anthropogenic greenhouse gas emissions, or human-caused climate emissions. See Le Treut et al., "Historical Overview." See also Gore, *An Inconvenient Truth*.

how "[p]eople in the rich world are increasingly concerned about emissions of greenhouse gases from developing countries" and it observes that "[t]hey tend to be less aware of their own place in the global distribution of CO_2 emissions."[35] The report estimates "the carbon footprint of the poorest 1 billion people on the planet at around 3 percent of the world's total footprint."[36]

The same report publishes a visually stunning cartogram that bloats nations responsible for a larger share of CO_2 emissions relative to those contributing a lesser share of emissions.[37] On that map, the African continent is nearly absent because its emissions are so low relative to the rest of the world. According to the International Energy Agency, more than 1.3 billion people are without access to electricity, the production of which contributes greatly to many nations' climate emissions. Most of the people who do not contribute to global emissions are unable to produce their own electricity, and more than 95 percent reside in sub-Saharan Africa or developing Asian countries.[38] The connection between relative global wealth or the "electricity haves" and their blameworthiness for human-induced climate change is as apparent as the geographical distribution of the "electricity have-nots" who suffer around the globe because of their poverty.[39]

The disparities surrounding climate change, and questions about who benefitted from causing climate change and who ought to pay for mitigation and adaptation going forward, are a key part of the ongoing challenges that hold up international negotiations. During the 15th Conference of the Parties (COP15), held by the United Nations Framework Convention on Climate Change (UNFCCC) in Copenhagen during 2009, much of the world had largely expected a binding international agreement on climate change. There was so much time and attention given to the negotiations of this forum that many in climate advocacy circles were astonished when no such agreement was reached in Copenhagen, or at each of the next three COPs I attended as an Observer Delegate with the Sierra Club. I was amazed by how so much popular momentum could have so little influence behind the closed doors of those backroom sessions to which only high-ranking government officials were privy.

35. UNDP, *Fighting Climate Change*, 43.
36. UNDP, *Fighting Climate Change*, 43.
37. UNDP, *Fighting Climate Change*, 42.
38. International Energy Agency, "Energy for all," 3.
39. Friedman, *Hot, Flat, and Crowded*, 63.

From my perspective, it felt as though something was corrupted about those negotiations. I had heard there was a persisting concern among some nations that issues of basic fairness, equity, and accountability were not given the attention they believed was necessary for a binding agreement to be considered fair for all signatory parties. Some argued that those countries that have historically profited the most from the causes of climate change, like the United States and much of Western Europe, ought to shoulder more of the burden of mitigation than others who have historically contributed fewer climate change emissions. Wealthy nations were not ready to agree to the proposed limitations on growth, the proposed creation of subsidy funds for poorer nations, and proposed reparations from wealthier countries to poorer countries in exchange for any caps to which they would agree.

While the worst effects of climate change bear down most forcefully upon vulnerable nations, more affluent and politically powerful nations tend to lack a sense of urgency and shared sacrifice necessary in addressing these challenges. Perhaps nothing illustrates this better than the disparity in access to clean drinking water. As water shortages become more frequent, they will crush the livelihoods of the world's poorest and become a heavy burden borne especially by women.[40] When water is already a precious commodity for many across the African continent, what will decreased access to water mean for millions of people, especially for a family whose mother already must spend several hours of her day procuring water for her household? When the nearest well runs dry, she will have even further to walk and her decisions in household management will become even more difficult as she prioritizes what can and cannot be done given the new limitations placed on such a precious resource many take for granted when they turn on the tap. Washing and cooking become even more time consuming, and it may become necessary that she keep a daughter home from school to help shoulder some of the burdens of these tasks.

The poorest people across the globe are also exposed to some of the most devastating diseases, even those that have generally been mitigated in more developed contexts. Shifts in the range and transmission potential of malaria are predicted across much of the African continent as the climate

---

40. Estimates for water shortages predict that between 75 million and 250 million people living on the African continent may be "exposed to increased water stress due to climate change" by 2020. See Parry et al., eds., *Contribution of Working Group II to the Fourth Assessment Report of the Intergovernmental Panel on Climate Change, 2007*, 13; See also Brown, *Plan B 4.0: Mobilizing to Save Civilization*, 38–48.

changes.[41] As malaria makes its way around the continent, moving in and out of various communities, its devastation is expected to intensify in both severity and scope.[42] New diseases will ravage communities wholly unprepared and unequipped to fight them.

As floods increase across much of South Asia, more people will die from diseases such as cholera. Although cholera is often associated with poor sanitation infrastructure, it can also occur when floodwaters mix contaminated sewage into drinking water.[43] Cholera is a terrible disease that can cause severe watery diarrhea and vomiting, possibly leading to death if untreated. Many people living in parts of South Asia are already too familiar with the disease since outbreaks there are beginning to both increase and intensify.[44]

Flooding also affects agricultural production and the availability of potable fresh water sources, especially when seawater is the source of floodwater contamination. According to one scenario presented by the World Bank, flooding from sea-level rise could inundate half the rice-growing land in Bangladesh—home to 160 million people.[45] Delta regions are historically home to several of the world's largest population centers. Those regions are particularly vulnerable to natural disasters due to environmental and climate changes, such as sea-level rise, storm surges, and river flooding.[46] Saltwater contamination of drinking water can force some to turn away from previously safe groundwater sources, and toward surface water sources not

---

41. Parry et al. eds., *Contribution of Working Group*, 12.

42. For analysis on the changes in disease transmission of malaria, see Gallup and Sachs, *Economic Burden of Malaria*, 1–22. See also Abeygunawardena et al., *Poverty and Climate Change*, 1–56.

43. A widely respected report finds that the "endemic morbidity and mortality due to diarrheal disease primarily associated with floods and droughts are expected to rise in East, South, and South-East Asia due to projected changes in the hydrological cycle." Parry et al., eds., *Contribution of Working Group*, 11.

44. An outbreak in the eastern Indian state of Orissa in 2007 killed at least 115 people and hospitalized more than 2,000. Bhubaneswar and Sanjaya, "Cholera Death Toll Rises." The causes of cholera outbreaks are complex, and usually related primarily to a lack of adequate infrastructure for both transporting and protecting potable water from sewage. Major flooding events, however, are a major cause of outbreaks because it disrupts whatever infrastructure may have been set in place for the separation of potable water from sewage water.

45. As cited in Brown, *Plan B 4.0*, 7; See also The World Bank, *World Development Report*; UNDP, *Human Development Report 2007/2008*, 100.

46. Parry et al., eds., *Contribution of Working Group*, 9.

yet contaminated by salt—but more likely to be contaminated by dangerous organisms and diseases, such as cholera. An individual without any access to water purification or testing technology can determine quickly and easily when water is salty. The presence of microbial contaminants is much more difficult to discern without advanced testing techniques.

Sometimes diseases like cholera are considered merely a development problem, something that proper infrastructure and sanitation will keep at bay. Climate change, however, acts like a curve ball to development planning, exacerbating poverty, and disrupting development with unexpected disasters and unanticipated extremes.[47] It overwhelms traditional infrastructure planning because the goalposts for what is acceptable keep moving, forcing communities to start over again whenever an unprecedented disaster occurs. Such fits and starts make it nearly impossible for some nations and many cities to claw their way out of extreme poverty.

It is both unfortunate and unfair that the heaviest burdens of climate change so disproportionately affect the poorest and most vulnerable people around the globe. All the while, the international community appears paralyzed by the challenge of climate change. The horrible problem of gender disparity and the tragedy of global poverty only complicate the enormity of the great work ahead of us. Still, we must have the courage to look at the world's suffering squarely in the face and respond to it with hope.

Even amidst the shadows and sorrow of such great despair, there is reason enough to act. Never before in our history has the human species had the ability to mobilize our collective creativity as a truly planetary civilization working together to solve common problems. We have grown into a geological force that has been mostly destructive until now, but we can choose to harness that energy to create a more just and humane world conducive to the flourishing of all life on Earth. I think wisdom embodied in the world's religions can be tremendously helpful in sustaining this task—one that is at root a process involving spiritual transformation, a deep imagination, and vision.

Religious stories have in the past been able to help people face incredibly daunting hardships. They can sometimes inspire us to move beyond the shortsighted and narrow self-interests that too often divide us, by uniting us in a shared effort to overcome adversity. The stories religions tell—their ability to shape human consciousness, to inspire and sustain us, and to mobilize us into action even amidst the depths of despair when struggles

---

47. See Shepherd et al., *Geography of Poverty*.

appear so insurmountable—this is the great hope embodied by a flickering flame burning boldly in a vast darkness. The work appears overwhelming, but the same power that ignited the universe into being still flows through it. It flows through all life forms on our planet and it courses through our veins too. In this struggle for life on Earth, perhaps this legacy is part of our light in the darkness.

3

A Faint Tracing on the Surface of Mystery[1]

I REACHED FOR MY small flashlight and quietly readied myself in the dark. The alarm had gone off at 5 AM, but given my hotel's isolated location and its limited access to fuel, the building's sole power generator had not yet been switched on. I had hoped to be long gone by the time it came online. It would be an hour-long hike to Tikal's tallest Mayan temple complex. That hike through the Guatemalan rainforest's seemingly impenetrable darkness ended up unlike anything I had ever experienced. It felt as though the night completely enveloped me while bloodcurdling screeches of howler monkeys screamed from every direction. When troops of howler monkeys call to each other in the night, to the human ear they sound eerily similar to the screams of young children in distress. Their howls were deafening.

By the time I made it to the temple, the soft pre-dawn light had already begun to quiet the monkeys. With sweaty anticipation and a pounding heart that felt as though it was about to leap from my chest, I hurried up the side of that Mesoamerican pyramid, Temple IV—a pre-Columbian structure a local guide later told me is the tallest still standing. Its pinnacle clears the rainforest canopy, and I had hoped to catch the sunrise over the horizon. From the top, I gazed eastward in complete silence with a few other souls who had made the trek. My breathing soon relaxed. For as far as the eye could see, individual trees and the fluttering of myriad life forms

---

1. Annie Dillard coins this phrase in *Pilgrim at Tinker Creek*, 11.

gradually became more pronounced. The mist slowly sank into the forest. Then, from atop that ancient Mayan temple, there was the light, piercing through the haze and emerging over the horizon. A chorus of birdsong wafted above the jungle. And so, the night gave way to morning's light, accompanied by a cacophonous celebration of creaturely elation. This was, for them and for me, the way it seemed every day should be greeted: with unrelenting hope and unbridled enthusiasm for the magnificent wonder and awe that each new day presents.

When I reflected later, I was surprised to discover that the sunrise itself was almost a non-event in comparison to the feelings awakened within as I remembered the way the forest appeared to miraculously come to life that morning. My Mayan guide had explained to me that the way popular culture had often sensationalized Mayan prophecies by depicting them in films as foreseeing the climactic demise of civilization as we know it, betrayed a fundamental misunderstanding of those stories. He explained that they were not about cataclysmic ends, but rather about the possibility of new life, and the struggle of birthing it into existence. The birthing process is not a warm and fuzzy transition like the caterpillar's metamorphosis into a butterfly. In human terms, birth is marked by hard, sweaty labor, often with the risk of death. The promise of new life is buried in that struggle, but the struggle makes new life possible. Sunrises, it seems, have this in common with birth: they create opportunities for profound transitions. I see my memory of that sunrise in the rainforest, and the daily awakening of life on earth, as a kind of metaphor for the various awakenings I hope are still possible in Christian thinking about who God might be and what it means to live a fully human life.

When Christians turn to their sacred stories and religious traditions, what are some of the images they might discover, rediscover, or even create anew that awaken their hearts and minds to the daily possibilities of a life well lived? Centuries ago, the Maya harnessed the power of story and cosmic thinking to move a people to build one of the most impressive civilizations history has ever known. The opportunity on the morning of this new day is to tap into that same kind of power to rebuild our civilizations so that they nurture the flourishing of all life.

The world's religions have an important role to play in that work. Christian traditions, so rich in symbols and rituals capable of inspiring the moral imagination, have a role too. There are moving narratives that can help reorient billions of the world's Christians with a sense of self that is grounded in a larger framework of creation, rather than a sense of self defined by social media, consumer electronics, and a ruthless "dog eat dog" economic mentality. There are images of the self, of the world, and of God that can shape and reshape the way many Christians see themselves in a world where everyone is asked to grapple with the very new challenge and possibility of planetary demise.[2] Here at the sunset of one era in human-Earth relations, is the promise of tomorrow presenting an incredible opportunity for humanity to birth itself into the dawning of an entirely new era.

Thousands of years of Christian history have borne witness to diverse images of what it means to be human in relation to others, God, and the world, including images that now appear to obstruct a more just and verdant future. This makes the work of responsibly sifting through all those ideas a difficult one, and any effort to do so will always be an unfinished work, constrained by the perspectives and social context of those engaged in it. Nonetheless, the retrieval and reconstruction of those wisdom traditions describing what it means to be human in relation to others and the world is a critically important task. So, this chapter begins with a recovery of some eclipsed images embedded in a long and diversiform Christian history of ideas. It then moves on to some more imaginative suggestions that are only now beginning to blossom in a garden bed tilled by the twin blades of ecological crisis and scientific insight.

In the following sections, so as not to get lost or overwhelmed, we will follow but a few threads of creation-centered streams of thought that I think can act as examples of opportunities for renewal in Christian faith. We will look specifically at accounts of theological anthropology and cosmology, or

---

2. From the perspective of thousands of years of Christian intellectual thought, contemporary challenges like climate change are relatively new. There are also new intellectual resources at our disposal to help humanity collectively grapple with such challenges, though I am not convinced it would be wise to utterly disregard the power and promise embodied in premodern ideas. A process of retrieval and appropriation of premodern ideas without regard for both the challenges and insights of the contemporary context would be as unfortunate as would a total disregard for the value and potential of ancient traditions to inform contemporary discourse. Elizabeth Johnson says it well when she observes that "[t]his is not to say that pre-modern Christian thought about creation led our ancestors to a highly developed ecological consciousness such as is needed today. Present wonder and protest at wasting the world reflect a genuinely new moment in history; our scientific knowledge, technology, and means of imaging and communication are genuinely different. But such theology kept alive the sense that creation had a certain religious value and was deserving of a modicum of respect when subject to human action." See Johnson, *Losing and Finding Creation*, 4.

what theologians call ideas about what it means to be human in relation to God and the universe. I hope to show what it looks like to comb through neglected but persistent ideas in Christian theology, critically evaluate them, and pull them forward in time.

There are themes within Christian traditions that can be drawn upon to frame the human person in a more humble and communitarian framework than the one that has come to predominate since the modern turn. Genesis 2 inspires one such theme that can help in the retrieval of pre-Christian Hebrew ideas of the human person as an earthen clay body. Another one frames the human person in terms of our residing within an inherently good cosmos and that image is a retrieval of Augustine of Hippo's contributions from the early Christian patristic period. We then progress into the medieval period and look to one of the period's most important theologians, Thomas Aquinas, to retrieve a theology of the common good relevant to contemporary issues. We attend to Christian theologies since the Reformation across both Roman Catholic and Protestant traditions as they have developed in the modern period, to paint a picture of what I suggest can be expressions of earthy sacramentalism and enfleshed spirituality for Christians today. The way we explore these ideas and pull them forward to serve as a kind of foundation for new theological ideas, demonstrates but one way to tap into the power and promise of Christian traditions at a time of ecological crisis. The goal is to inspire Christians to creatively harness stories that matter and employ them in ways that contribute to a more equitable, flourishing world.

EARTHEN CLAY BODIES

Some of the most moving, persistent stories binding communities together are those shaped over generations, spanning great lengths of time and space in their telling and retelling. Even when they appear to cease speaking to the needs or concerns of a people, they may linger on in the collective consciousness of a community by subtly influencing those stories that do occupy a more prominent role. The biblical creation stories found in Genesis 1 and 2 probably have long resided there in the background, covertly shaping aspects of modern worldviews in less than obvious, but no less important ways. A close look at the themes contained in these stories is warranted.

In the creation story of Genesis 1, for example, there is a theme of human dominion over the Earth—one that describes humanity's special role within creation as that of subduing all the rest. When God creates humankind in Genesis 1, people are created in God's image, and instructed to "[. . .] fill the earth and subdue it; and have dominion over the fish of the sea and over the birds of the air and over every living thing that moves upon the earth" (Gen 1:28). The text also presents a picture of an all-powerful God, and a cosmos in which God simply speaks creation into being and it is so. The story emphasizes an innately transcendent quality of the creator, and it offers a vision of humanity shaped in that same world-transcending "image" or "likeness" of God (Gen 1:26).

In the story of creation in Genesis 2, humanity is molded from the Earth, breathed into life by the same force that animates all other breathing creatures (Gen 2:7), and is commissioned to till and keep the land (Gen 2:15). Genesis 2 emphasizes the earthiness of humanity as a core part of what it means to be human. It is that part that connects human beings to all other creatures shaped from and for the Earth. It also offers a contrasting vision of how people might imagine themselves in relation to others and before God.[3] In Genesis 2, God breathes, forms, and shapes while people are made, loved, and connected—not in an abstract way, but intimately so. Each of the Genesis stories present a clear and compelling image of God, as well as a narrative about what it means to be human before God and in relation to others—but those narratives are told from perspectives that differ profoundly.

Close, careful readings of the Genesis narratives by biblical scholars such as Theodore Hiebert offer penetrating insights into early Israel's cultural context and can help explain why these two stories are so different.[4]

---

3. See Hiebert, *Yahwist's Landscape*. See also Hiebert, "Human Vocation," 138–41.

4. Theodore Hiebert is an influential biblical scholar and expert in classical Hebrew language as well as Hebrew Scriptures. He offers a careful examination of the oldest narrative sections of Genesis and specifically those sections generally attributed to what is called the "Yahwist" or "J" writer, and he holds those Yahwist accounts in tension with what are called the "Priestly" accounts of creation in Genesis. See Hiebert, *Yahwist's Landscape*, 15. While the authorship of Genesis is debated, there is significant scholarly consensus concerning the text's development over time and authorship by multiple authors. It is argued, for example, that "[a]s a result of hundreds of years of scholarly analysis, we now know that the book was written over centuries by multiple authors, and we have a relatively specific and assured picture of the final stages of its composition (the combination of P with non-Priestly materials). These findings highlight the way Genesis is not limited to just one situation or set of perspectives. Instead, it is a chorale of different voices, a distillate of ancient Israel's experiences with God over the centuries, written in the form of continually adapted stories about beginnings." See Carr, "Genesis," 327. See also Newsom and Ringe, *Women's Bible Commentary*, 14–17.

The Genesis accounts of creation are sometimes casually interpreted as a single story, but though they share similarities they really are so different that it is important to discuss them separately. They contain two distinct accounts of creation with highly contrasting images of what it means to be human in relation to the rest of creation, and this represents the worldviews of storytellers from vastly different time periods with different priorities and experiences of the world.[5] Whereas Genesis 1 betrays an ancient priestly perspective rooted in structures of hierarchal power and authority, Genesis 2 appears to have emerged from the life experiences of common "subsistence farmer[s] in the Mediterranean highlands."[6] The two accounts in Genesis, as Hiebert describes them, are actually "almost inverse images of one another."[7] Hiebert notes that in the Genesis 1 account, "the human [is] created alone in God's image, as distinct from other forms of life" while in the Genesis 2 account humanity is "made like the animals from the arable soil, as related to other forms of life."[8] In Genesis 1, humanity is imagined "as master of the earth," and "the human vocation is one of dominion and supervision," but in Genesis 2 human vocation is imagined as one of "dependence" upon the Earth and "service" toward it.[9]

The mandate, in Genesis 1:28, for humanity to "fill the earth and subdue it; and have dominion over [. . .] every living thing" is one that is in keeping with the way its priestly authors probably would have understood the natural and inherent structure of their society and world. They likely would have held a hierarchical view of the way nature and society ought to be rightly organized—a social hierarchy in which they would have had a place very near the top and one that they would have considered essential to the maintenance of early Israel and to have been ordained by God.[10] Conversely, the context of those imagining and telling the creation story

---

5. Hiebert, "The Human Vocation," 136.

6. Hiebert, "The Human Vocation," 139. Also Newsom and Ringe, *Women's Bible Commentary*, 16. While I argue here that it is important to reclaim some of the lost voices within the Genesis narrative, I share Larry Rasmussen's sense of loss for the "massive deficit" garnered by the loss of those pre-Neolithic perspectives generally found lacking in the creation stories recorded in the faiths of Peoples of the Book. See, for example, Rasmussen, *Earth-Honoring Faith*, 52.

7. Hiebert, "The Human Vocation," 140–41.

8. Hiebert, "The Human Vocation," 140–41.

9. Hiebert, "The Human Vocation," 140–41.

10. Hiebert, "The Human Vocation," 136–38. As per Hiebert, see also Cross, " Priestly Houses," 195–215, especially 211, 215; Coote and Ord, *In the Beginning*, 29–56.

recorded in Genesis 2 is equally influential on the way the story develops. It reflects the life experiences and perspectives of a people who are wholly dependent upon the land and its cultivation for their way of life. As an agrarian people, they very likely would have had a close working relationship with the land and its processes, as well as an awareness of their own dependence on it for continued sustenance and survival.

The subsistence farmer's image of what it means to live out a fully human life is one that is profoundly more humble in its orientation to the other forms of life with which our species shares this planet.[11] The process of translation has covered over this aspect of the story. The verb used in Genesis 2:15 for the mandate "to till" the earth, ʿābad, may be more appropriately translated as "to serve." It is a word that in the Hebrew scriptures, as Hiebert illustrates, "is the customary term to express servitude, of slave to master (Gen. 12:6), of one people to another (Exod. 5:9), and of Israel's service to God in its life and worship (Exod. 4:23)."[12] Instead of reading God's instruction to humanity as an agrarian mandate to till the earth, it may be more helpful to read it as a mandate to serve and keep the land. Moreover, the Hebrew word used to refer to human beings (ʿādām) reveals not only an etymological connection to the land (ʿădāmâ) from and for which humanity was made, but also it is a word that may have been more instantly recognized as connoting the grafted nature of humanity's symbolic connection to, and dependence upon, the earth.[13]

---

11. It should be noted that there is a tradition of interpreting the "dominion" passages in the Hebrew Scriptures, christologically. As Larry Rasmussen points out, this theme plays out strongly in some Christian stewardship models of creation care, since one perspective is that "[i]f Jesus is dominus (Lord), then the human exercise of power should be patterned on his kind of lordship—a servant stance in which the last are made first, the weak are made strong, and even the sparrow is cherished." See Rasmussen, *Earth Community, Earth Ethics*, 231. As per Rasmussen, see also Hall, *The Steward*; and Hall, *Professing the Faith*. For other accounts in the scriptures that position the human in terms of a framework of planetary humility, see Bouma-Prediger, *For the Beauty*, 146. And McKibben, *Comforting Whirlwind*, 34. For more on these passages from an expressly Jewish perspective, see Lerner, *Jewish Renewal*, 416.

12. Hiebert, "The Human Vocation," 140.

13. Hiebert, *The Yahwist's Landscape*, 34–36. Here I choose to refer to ʿādām with the more inclusive "human beings" instead of "man." As Hiebert notes, there is not yet consensus as to whether "the first person created in this account is male, sexually undifferentiated, or androgynous." See Hiebert, "The Human Vocation," 139, 152. See also Olson, "Untying the Knot," 73–88, especially 76–78.

A Faint Tracing on the Surface of Mystery

Until the election of Pope Francis and the release of his encyclical letter, *Laudato Si'*, Genesis 2 had been mostly ignored.[14] The Vatican has had a longstanding tendency of referencing Genesis 1 almost exclusively wherever the topic of creation is concerned.[15] With few exceptions, other Christian communities have not done much better. This is a most unfortunate oversight at a time when images of humanity's embodied earthiness and connectedness with all other life on Earth can offer a much more constructive framework for human-Earth relations than one in which we are framed as subjugators of a dominion. To talk about the Earth in a way that reduces it to a collection of objects to be refashioned for human use is to amputate the power and promise of those traditions at a time when human civilizations have objectified creation to near oblivion.[16]

The appropriate question is not really whether Christian Scriptures paint a picture of humanity as ecologically embedded or estranged, for both images are present in the traditions. The question is rather: from which set of stories ought people of faith to draw inspiration in directing their general orientation toward others and the world? For Christians, this means either cherry picking passages to suit one's preferences or viewing all of Scriptures through a lens emphasizing Christ's liberating message of love and justice. A great opportunity for Christians today is to reject a vision of the Earth as a place to be dominated, subdued, and managed exclusively for the flourishing of some while others perish. There is a choice to instead see and approach the world in terms of the gospel call and as a place to be tended for the flourishing of all living beings.

The Hebrew words used in Genesis 2:7 describe with penetrating symbolism the way its storytellers imagined humanity coming into being the moment God breathed life into our earthen clay forms. In that instant, human beings joined a family of breathing creatures that God had already created—we became nepeš ḥayyâ, which in Genesis 2:19 is the phrase used in common for *all* breathing creatures, from the birds of the air to the animals of the field.[17] That is a hopeful image for a broken world. Birds, beasts,

---

14. Francis, *Laudato Si'*.

15. French, "Greening *Gaudium Et Spes*," 200–201.

16. As Pope John Paul II wrote, "through work man not only transforms nature, adapting it to his own needs, but he also achieves fulfillment as a human being and indeed, in a sense, becomes 'more a human being.'" Pope John Paul II, *Laborem Exercens*, paragraph 9.

17. Hiebert, "The Human Vocation," 139. See also Hiebert, *The Yahwist's Landscape*, 63; and Anderson, *From Creation*, 157.

and human beings all draw a breath of life that animates their being and connects them to every other being that people of faith believe God has called into existence.

A GOOD COSMOS

Some of the earliest Christians would have understood the Hebrew creation stories as affirming the inherent goodness of creation. That view of the world as inherently good was not without its critics, however, and various philosophies and theologies since then have on occasion helped to nurture an ongoing antagonistic view of the material world. One important Christian theologian, who vigorously defended the idea of creation's inherent goodness during the earliest days of the Christian church, was Augustine of Hippo. He was born in 354 in North Africa (or modern-day Algeria) and died from fever in the midst of a city under siege in 430.[18] Augustine's written body of work is quite substantial and wide-ranging with regard to both the breadth and depth of topics he covered. He is rightly credited with helping to settle a number of debates that emerged during what is now called the Patristic Period—a time when Christian beliefs and ideas were becoming increasingly unified and systematized. His defense of the inherent goodness of creation—a general view of the world that may very well prove helpful during this time of ecological crisis—has too often been eclipsed or confused by his many other teachings.

Theologians since Augustine have frequently been consumed by an enduring focus on his anti-Pelagian writings regarding the concept of original sin. In addition to a heavy focus on sin, there has been a focus on Augustine's Platonic and Neoplatonic dualisms that continue to play a predominating role in some Christian theologies today, even while his teachings on the goodness of creation are often flatly ignored. Augustine's thinking certainly does allow that the material world, including human bodies, may become corrupted through what he considered to be the evils of sin.[19] Still, sin is not something Augustine described as originating in the beings of this world or in human bodies inherently. When Augustine reflected on his own sin, he concluded that it consisted in his seeking "pleasure, sublimity, and truth not in God but in his creatures, in [himself] and

---

18. See Mahoney, *Making of Moral Theology*, 37–70, especially 39–40. See also Brown, *Augustine of Hippo*, 7, 430.

19. Regarding corruptibility of the body, see Augustine, *Concerning the City*, 548–52.

other created beings."²⁰ Augustine's emphasis is on the *turn away* from God rather than in any inherent kind of evil in such things.

For Augustine, sin is in the messed up order of desire, and the effort to look for ultimate fulfillment in places he does not think it can be found, which is to say, outside of God. The world, as Augustine described it, is a good and ordered place created, animated, and sustained by God's vivifying power. Christians are called to seek God in the right ordering of their desires, their time, their attention, and their focus. Sin messes all that up—it corrupts things by distracting Christians from God's goodness. Sin turns Christians away from God, but everything God created is good by the very fact of its existence. The world is a good creation.

Augustine's writings assert his belief that all things are originated from and dependent upon God. Everything God created is inherently good by virtue of it being called forth into existence at the behest of a good creator. As Jame Schaefer points out in her analysis of patristic and medieval texts, Augustine held a pervasive view of the goodness of creation.²¹ "The earth is good by the height of its mountains," he wrote, "and the evenness of its fields."²² Even in the most mundane aspects of our lives, "good is the food that is pleasant and conducive to health; and good is health without pains and weariness; and good is [. . .] the soul of a friend with the sweetness of concord and the fidelity of love [. . .]."²³ The very act of being, or simply existing, is akin to,

---

20. Augustine, *Confessions*, Book I, Chapter xx. See also Taylor, *Sources of the Self*, 136. On the importance of the "turning" aspect of Augustine's thought, Taylor observes that "Augustine takes our focus off the objects reason knows, the field of the Ideas, and directs it onto the activity of striving to know which each of us carries on; and he makes us aware of this in a first-person perspective. At the end of this road we see that God's is the power sustaining and directing this activity."

21. Schaefer, *Theological Foundations*, 18. Among the several examples to which she points, the following is perhaps one of Augustine's most poetic and moving. He says, "The earth is good by the height of its mountains, the moderate elevation of its hills, and the evenness of its fields; and good is the farm that is pleasant and fertile; and good is the house that is arranged throughout in symmetrical proportions and is spacious and bright; and good are the animals, animate bodies; and good is the mild and salubrious air; and good is the food that is pleasant and conducive to health; and good is health without pains and weariness; and good is the countenance of man with regular features, a cheerful expression, and a glowing color; and good is the soul of a friend with the sweetness of concord and the fidelity of love; and good is the just man; and good are riches because they readily assist us; and good is the heaven with its own sun, moon, and stars." See Augustine, *The Trinity*, 274. As cited in Schaefer, *Theological Foundations*, 18–19.

22. Augustine, *The Trinity*, 274. As cited in Schaefer, *Theological Foundations*, 18–19.

23. Augustine, *The Trinity*, 274. As cited in Schaefer, *Theological Foundations*, 18–19.

and even interchangeable with, the way Augustine conceived of goodness. He argued quite directly, "as long as [things] exist, they are good."[24] In the same passage, he repeats the statement and resolves "[f]or our God has made 'all things very good' (Gen. 1:31)."[25] Augustine seems to see God's declaration in biblical Scriptures as grounding his own conclusion in a right interpretation of theology. Augustine's defense of the world's inherent goodness stood out to him as a marker of Christian orthodoxy. Moreover, this belief eventually came to mark a distinction between Christian worldviews and those views that Augustine vehemently resisted in the competing, popular philosophies increasingly widespread during his time.

Augustine's unique background prepared him to argue particularly well on several of the major disputes arising between Christian communities and competing ideologies. One dispute had to do with what is now called metaphysical dualism, or the way the cosmos might be imagined as comprised of two fundamentally different and opposing forces, such as good and evil or spirit and matter.[26] According to some of the prevailing belief systems of his day, such as Gnosticism and Manichaeism, that view of the world saw matter as evil and as corrupting the spirit. Before Augustine was baptized a Christian in 387, Augustine was initially attracted to Manichaean ideas, and he knew them quite well.[27] When he was later ordained a Christian priest and then bishop of Hippo in 395, Augustine "spent the rest of his life, presiding, arbitrating, conferring, debating, preaching, writing, and above all, attacking the Christian deviations of Donatism, Manichaeism, Arianism, and especially, Pelagianism."[28] These belief systems and their views of the world competed against Christianity for adherents during the development of the early Christian church.[29] Augustine's

---

24. Augustine, *Confessions*, 124–25. See also Augustine, *Concerning the City*, 473.

25. Augustine, *Confessions*, 124–25. See also Augustine, *Concerning the City*, 473.

26. G. R. Evans notes in his introduction that such dualisms also regarded the Manichaeist and Gnostic idea that "there are two powers in the universe, two 'first principles', good and evil, eternally at war" which he argues is "ultimately incompatible with the Christian belief in one God, who is omnipotent and wholly good . . . ." See Augustine, *Concerning the City*, xxiv.

27. He was also attracted to Donatism. See Forell, *History of Christian Ethics*, 155–57.

28. Mahoney, *Making of Moral Theology*, 39.

29. Mahoney, *Making of Moral Theology*, 38. As per Mahoney, see also Brown, *Augustine of Hippo*, 46–60; and Bonner, *St. Augustine of Hippo*, 157–236. For an account of Manichaeism as it is distinguished from Gnosticism, see BeDuhn, *Manichaean Body*. For an exploration on Augustine's dissatisfaction with Manichaeism, see BeDuhn, *Augustine's*

intimate familiarity and disillusionment with Manichaeism, and later his lifelong resistance to its ideals, allowed him to make a persuasive case for a Christian affirmation of creation's inherent goodness.

The "Gnostic problem," or rather the early church's "contact and conflict" with Gnosticism during the second to fifth centuries, proved to be a particularly competitive framework.[30] On the one hand, Gnosticism asserted a vision of the world in which "people live in the world, but at least some of them are not of it."[31] Gnosticism viewed human bodies as "made of the same substance as the world," but argued that "there is, or may be, hidden within each one a spark of the divine life" that would have been considered wholly spiritual.[32] Salvation, then, would have been regarded as "the rescuing of the divine spark from its imprisonment in the material world, and specifically in the flesh."[33] Everything physical, including human bodies, is perceived to be intrinsically evil.[34] Augustine defends Christian thinking against such a view. So even while Augustine's writing does maintain some Platonic dualisms between body and soul, he is very clear in the lattermost developments of his thoughts that any notions of matter and body as intrinsically evil are incompatible with the belief that God created all things to be naturally good.[35]

To reclaim Augustine's image of a good cosmos is to revive an ancient view of a good earth, animated and sustained by God's vivifying power. It can offer a view of the world as a place where Christian seekers nurture their search for God's goodness in the world, in the right ordering of their desire to see God's mercy and justice lived out *in* the world. If the right ordering of one's inclinations can be reimagined as seeking God in solidarity with others for the flourishing of all life on earth, then there may be much merit in retrieving this stream of thought. Moreover, if Christians can imagine God as the vivifying power that animates, sustains, and directs all life and

---

Manichaean Dilemma.

30. Spivey et al., *Anatomy of the New*, 42–43.

31. Spivey et al., *Anatomy of the New*, 42–43.

32. Spivey et al., *Anatomy of the New*, 42–43.

33. Spivey et al., *Anatomy of the New*, 42–43.

34. For a few examples, see Augustine, *Concerning the City*, 473–77, 568–71, 870–72. For a helpful overview of the shift in Augustine's thought on natural entities, see Ledoux, "Green Augustine," 331–44.

35. Augustine rejects the Platonic theory of body and soul outright in *City of God*. Regarding the eventual rejection of this aspect of Platonic theory, see especially Augustine, *Concerning the City*, 554–55.

not simply human life, then they may yet realize a way of seeing themselves, God, and the world in a way that is not only inspired by ancient Christian ideals but also ready to be a positive force for change in the world today.[36]

THE COMMON GOOD

As Christians assess their views on God and the world in light of ecological and social justice concerns, there is also a need to assess those relations among individuals and between them and the social systems people create. What are appropriately sustainable and just systems of governance and land ownership? How do Christians and their faith communities contribute to these systems in ways that benefit all people and all forms of life on Earth? Asked differently, what theological images can Christians draw upon to nourish a spirit of solidarity that empowers them to work together with all other people of good will for the common good? In this section, we turn to the teachings of Thomas Aquinas to recover an image of God who calls and binds people together in the shared use of resources for the common good. It builds on the directional emphasis of Augustine's theology, to imagine an active role for Christians in building the future City of God in the here and now.

Thomistic theology offers an inherently social perspective on human nature, but it needs some updating to be refreshed for the twenty-first century. Thomas Aquinas was born in 1225 and admitted into the faculty of theology at the University of Paris in 1257 at a time when "theology" was considered "queen of the faculties," and he had a highly intellectual and productive career before his death in 1274.[37] He lived during the high Middle Ages—a relatively prosperous time across Europe, sandwiched between the population declines and social turbulence of the early Middle

---

36. Charles Taylor says, "The clear difference between Augustine's imagery and Plato's, for all the continuity of metaphysical theory, reposes on a major difference of doctrine. Augustine takes our focus off the objects reason knows, the field of the Ideas, and directs it onto the activity of striving to know which each of us carries on; and he makes us aware of this in a first-person perspective. At the end of this road we see that God's is the power sustaining and directing this activity. We grasp the intelligible not just because our soul's eye is directed to it but primarily because we are directed by the Master within." He continues, "God can be thought of as the most fundamental ordering principle in me. As the soul animates the body, so God does the soul. He vivifies it." See Taylor, *Sources of the Self*, 136.

37. Baldwin, *Scholastic Culture*, 1, 79, 95. For a helpful overview of his life and published works, see also Sigmund, ed., *St. Thomas Aquinas*, xiii–xxvii.

Ages ushered in by the fall of the Western Roman Empire but preceding the calamitous Black Death so characteristic of the late Middle Ages.[38] The shift from traditions of empire and rulership to codified laws on rights and property was part of the Thomistic social context that influenced the trajectory of his work during the medieval period.

Contemporary Christians may find several of his ideas worth being pulled forward into the modern world, so long as they can be appropriately untangled from less helpful aspects of his historical context. Thomas drew upon the best science of his day to inform and inspire his theology, so for contemporary Christians to do likewise would be in keeping with the spirit of the way Thomas practiced theology. Most philosophers of his day would have accepted a traditional view of the world as organized according to a strict hierarchy or great chain of being with God at the top, followed by angels and people, with kings above commoners for example, and then below us eventually animals, then plants, and minerals near the bottom.[39] This now outdated image of the world was part of his social context but the science of our day offers an ecological image that is more of a web or bush than a chain. While the givenness of a natural hierarchy in human relations persists in various forms all over the world even today, it is rightly challenged in many places too. It may be fair to criticize retrospectively aspects of Thomistic theology as overly hierarchical in nature and generally anthropocentric in its orientation to the world. It is a privilege to be able to look back on history critically, with the benefit of hindsight and some distance from the prevailing influences of a particular period.

Despite the philosophical and scientific headwinds of Thomas's day, he offered an image of the human person as highly social in nature. He saw humanity as embedded within a larger cosmic framework, so his view of the human person is contextualized in a way that is far less antagonistic to an ecological worldview than may at first appear. "Aquinas balances," as William C. French argues, "an appreciation for the distinctiveness and worth of human reason, freedom, and agency with an appreciation for our

---

38. Summarizing some of the consequences of the medieval social context and their influence on intellectual thought, Santmire notes themes of "pervasive alienation from nature [characterizing] the early Middle Ages, and then the remarkable rebirth of interest in, and appreciation for, nature in the twelfth century." He argues this played a role in the development of a "theological naturalism" in Thomistic thought. See Santmire, *Travail of Nature*, 76, 75–95.

39. Lovejoy, *Great Chain of Being*.

embodied participation in the vast sphere of creation."[40] While acknowledging the human-centered focus found in Thomistic thought, Daniel P. Scheid argues that there is nonetheless much else about Thomistic theology that can be helpful in correlating the human common good with the planetary common good. Like French, Scheid argues persuasively that Thomistic thought offers a cosmic frame, emphasizing how "the whole universe surpasses in excellence any individual creature," including human beings, and "the order and diversity of creatures is in fact the best aspect of creation."[41] So while the strict hierarchy and generally human-centered focus of Thomistic theology unsurprisingly reflect some of the many limitations imposed by the prevailing perspectives of his day, this should not keep the larger cosmic framework of Thomistic thought from becoming a source of inspiration for contemporary Christians today.

Drawing on the best science of the day to inform theology is a central part of Thomistic theology. Following in his footsteps, it would be necessary to consider insights from the social sciences, ecology, and evolutionary biology when reclaiming Thomistic theology on the human person and in envisioning humanity's right relations with the world.[42] Those sciences

---

40. William French offers a stunning summary of this important nuance in Thomistic thought, and it is worth noting it in full: "Thomas supports a robust anthropocentrism that concentrates attention on humanity's unique creation as a rational being in the embodied world and it is this capacity for rationality, this distinctive intellectual soul, which humanity shares only with angels, that sharply demarcates humans as separate from and categorically superior to the rest of the embodied created world. Humans are at the top of the great scale of embodied being and thus are said by Thomas to be at the top of the corresponding scale of value. [. . .] But while Thomas develops a vigorous anthropocentric value scheme that celebrates the superiority of the human, he locates an understanding of the human within a cosmological frame that highlights that humanity is a participant in the broader community of the universe. [. . .] This is perhaps his most distinctive contribution for our reflection today. Where dominant streams of modern Protestant and Catholic theology, and indeed modern Western philosophy, have concentrated on an understanding of history as the dynamic frame for understanding human life, Aquinas balances an appreciation for the distinctiveness and worth of human reason, freedom, and agency with an appreciation for our embodied participation in the vast sphere of creation." See French, "Grace is Everywhere," 151–52.

41. Regarding Thomistic theology, Daniel Scheid argues that though Aquinas "certainly affirms humanity's dominion over non-human creation and its privileged position over all other Earthly creatures, he also envisions a cosmos in which all creatures, including human beings, contribute to a glorious cosmic end centered on God." See Scheid, "Thomas Aquinas," 127, 134.

42. I am thinking specifically, here, of the hard and precise cleavage between human beings and the rest of the world that is present in Thomistic accounts of the human

clearly show that humanity's capacity to survive and thrive is increasingly dependent on the ability of ecosystems to function and maintain the kind of ecological stability that is amenable to the support of all life. This also means that modern democratic principles of fairness and equity, as well as contemporary understandings of gender, the body, and sexuality, ought to serve as a lens through which Thomistic ideas are mediated. Serving the greatest common good requires, more than ever, a reappraisal of the way communities regard human equity and social justice on the one hand with the right use of property, and a deconstruction of the unfair, unsustainable practices of production and consumption on the other.[43] It is very much in keeping with the spirit of Thomistic theology for contemporary Christians to pull forward prominent streams of pre-modern Christian thought when they square with the best science of our day, like the Thomistic view of the human person as intrinsically social and so then inherently bound by duty to serve the common good.[44]

Thomas Aquinas taught that a person "should not possess external things as his alone but for the community, so that he is ready to share them with others in cases of necessity."[45] He did not argue against the idea of private property. He allowed that "it is legitimate for a man to possess

---

person and the world that often encourages and sustains a perspective about nature that voids a healthy sense of its moral worth. While I do not think that the distinctions Aquinas makes between people and other creatures create an insurmountable challenge for those seeking to reclaim the creation-centered framework of his theology in environmentally responsible ways, particularly as French and Scheid note, I do wonder if such a bifurcation makes it hard to "square the circle" with regard to what can be learned from ecological and evolutionary biology. Francisco Benzoni engages in such a project. See Benzoni, *Ecological Ethics*, 4–5. Also, using the best science of the day to inform theology is a central part of Thomistic theology, as French argues and as I soon note.

43. For Aquinas, the greatest common good is the good of the universe. See Aquinas, *Summa Theologiae*, 1.22.4, 1.47.1, and 1.103.2.

44. As French asserts: "One of the most impressive features of Thomas's work was his concern to correlate his received theological tradition and its acts of affirmations and perspectives with the best available science and understanding of the universe of his day. In his day, this was found in the newly available translations displaying the power and sweep of the Aristotelian world-picture. But to keep faith with Thomas's spirit and historical example, it seems best not to reify his assertions that are grounded in the thirteenth century's science and metaphysics, but rather to follow in his open engagement with the best of today's science to discern how creation, providence, and redemption might best be understood in light of the challenges and knowledge of the twenty-first century." See French, *Grace is Everywhere*," 164.

45. Aquinas, *Summa Theologiae*, 2.66.2. See Sigmund, ed., *St. Thomas*, 72.

private property; indeed" he argued, "it is necessary for human life [. . .]."[46] The key here is the way a Thomistic understanding of the human person is generally regarded. Individuals are obligated to manage those things under their charge. People are called to share and use their resources in a way that honors the bonded nature of human beings in their shared humanity before God. Humanity's fundamentally social nature bears consequent responsibilities and obligations to a concern and practice of care for others. The way people live and act in the world ought to flow from such ecologically grounded images as these.

The use of private property within a framework of individualism and autonomy, however, has flourished in much of the west during the modern era. In the United States during the nineteenth and twentieth centuries, some of the greatest wealth amassed by private individuals and their companies was through the exploitation of shared common spaces, such as forests for logging, land for mining, and both air and water as sinks for the waste products of industrial production. Those with the most capital to purchase property and finance the exploitation of natural resources saw the greatest gains with dramatic negative consequences for air quality, water quality, and biodiversity loss across all sorts of ecosystems.

The use of private property within a Thomistic framework emphasizes the way it is able to contribute to the larger common good. Thomas argued that all things exist for a larger good, and working toward the common good is how people contribute to what he described as the ongoing perfection of the universe.[47] This orients the human person in terms of our agency and responsibility to actively engage the world and make it a better place. The collective use of property during the modern era, and in particular the use and abuse of land, water, and biosphere has resulted in part from a deeply degraded view of the human person. In that view, the individual is

---

46. Aquinas, *Summa Theologiae*. Sigmund, ed., *St. Thomas*. Aquinas lists three reasons why he does not argue against the concept of private property and rather for its right use. He argued that with private property, "everyone is more concerned to take care of something that belongs only to him [. . . that], human affairs are more efficiently organized [. . . and] peace is better preserved [. . .]." Aquinas, *Summa Theologiae*, 2.66.2. See Sigmund, *St. Thomas*, 72.

47. While Thomistic theology is sometimes perceived as anthropocentric in the sense that Aquinas saw nonhuman beings as being created for the good of human beings, it is also important to point out that Aquinas also recognized intrinsic value in "nature" and that he held an overarching organic worldview in which humans existed for the good and perfection of the larger universe. For a helpful explication of this argument, see LeBlanc, "Eco-Thomism," 293–306. See also Jenkins, *Ecologies of Grace*, 118–21.

divorced from the larger privilege and responsibility of creating a world in which all ought to have an opportunity to not only survive but to thrive and flourish while caring for our common home. A Thomistic theology of the person frames our humanity in terms of a network of relations to property, resources, and their management in solidarity for the common good.

A Thomistic sense of the common good, as Susanne M. DeCrane observes, can be framed in terms where "God is the ultimate common good of all creation," but she notes that the "common good is also understood by Aquinas as being connected to the practical exigencies of living in society."[48] Said differently, "[h]ow we live in society, how we shape our societies and our relationships within societies, is related to the pursuit of God as the highest good."[49] These are not wholly abstract images of what it means to be a human person and live a good life. They are practical; they are ready to be employed by everyday people as they go about the messy business of figuring out what it means to live a fully human life in a complex world. To be human "all in" is to recognize the way individual human flourishing is so often quite dramatically bound up with the flourishing of others. Owning property, managing the resources at our disposal, producing, and consuming goods are all opportunities for each of us to participate in, and recognize the way we value—or do not—relationships with others. It is how any one person can have a role in actively cultivating the kind of society that invokes and constitutes the cosmic common good.[50]

---

48. DeCrane, *Aquinas, Feminism*, 60. Here, DeCrane notes with impressive articulation, that "[t]he central importance of the common good in Aquinas's moral thought flows naturally from his anthropology. As intrinsically social beings, we exist and flourish only within the context of a community. Therefore, Aquinas writes that because of our social nature, we are obligated to do 'whatever is necessary for the preservation of human society.' The issue is not merely the preservation of the sheer existence of the group, but of its flourishing as a necessary, life-promoting reality for all members of the society." DeCrane, *Aquinas, Feminism*, 59–60. As per DeCrane, see also Aquinas, *Summa Theologica*, 1.96.4, 2.47.10, 2.109.3; Aquinas, *On Kingship*; Scully, "Place of the State," 107–29.

49. DeCrane, *Aquinas, Feminism*, 60.

50. The common good is so important to Aquinas's theology that Jean Porter argues: "Aquinas insists, as strongly as any Marxist, that the common good takes precedence over the good of the individual, just as the good of the universe as a whole is a greater good than the good of any one creature, however exalted [that creature] (2.47.10; 2.58.12; 2.64.2). And yet, Aquinas is not in fact the one-sided communalist that these remarks, taken alone, would suggest." Though Porter does parallel this particular thought of Aquinas to Karl Marx, DeCrane continues in her text to nuance Porter's statement by noting that a key difference between Marx and Aquinas is that, for Aquinas, "the common good

Thomas Aquinas, in some ways much like Augustine, offers images of the human person and the world in which both have value before God. We share a common home with an unimaginably vast array of other life-forms as part of a mutually interdependent web of life that sustains and supports the shared flourishing of all life on this planet. Our corner of the cosmos is not a dead, barren, and lifeless place, even if vast stretches of the universe appear to be so. From the vantage point of Earth, people of faith can imagine the cosmos as a place in which "the love of God infuses and creates goodness in things."[51] Aquinas, like Augustine, believed ours is a good cosmos, and he believed that "because [God's] goodness could not be adequately represented by one creature alone, [God] produced many and diverse creatures, so that what was wanting to one in the representation of the divine goodness might be supplied by another."[52] He believed that "the whole universe together participates in the divine goodness more perfectly, and represents it better, than any given single creature."[53] Our good cosmos, as Christians are invited to imagine, is a place where God has bound us up and co-missioned us in the sacred task of incarnating solidarity with other earthen clay bodies in shared pursuit of the common good.

EARTHY SACRAMENTALISM

Christian communities have a legacy of cultivating their heritage so the limbs of tradition are able to sprout fruit under entirely new conditions. Christian theology, rituals, and traditions have never been static—they grow in response to a life lived in faith as the church grows, or at least they should—and this dynamism is part of what breathes new life into the church, continuing its relevance in the modern world. Pope John Paul II writes that the church's social teachings are to be perceived as an "*application* of the word of God to people's lives and the life of society, as well as to the earthly realities connected with them."[54] This means that a central task for those in the church today is to ask how Christian ideas can be

---

in itself is privileged for the sake of promoting and producing the circumstances that will aid the members of the community to grow in their individual goodness and happiness." See Porter, *Recovery of Virtue*, 125; As cited in DeCrane, *Aquinas, Feminism*, 73, 74.

51. Aquinas, *Summa Theologiae*, 1.20.2.
52. Aquinas, *Summa Theologiae*, 1.47.1.
53. Aquinas, *Summa Theologiae*, 1.47.1.
54. Pope John Paul II, "Sollicitudo Rei Socialis," paragraph 8.

"lived out" in the religious life of communities.[55] Creative engagement with sacramental rituals is one way to enrich the experience of religious life within living faith communities. The impressive cultural power of religious ideas and the inventiveness of a strong moral imagination sustain my hope that Christian communities are able to awaken a new kind of earthy sacramentalism in the life of the church from the annals of tradition.[56]

Both Roman Catholic and some Protestant Christians maintain a belief that God's real presence can be experienced through bread and wine. Unfortunately, the way many Christians approach the communion table and reflect on Eucharistic traditions and rituals simply do not reflect a robust earthy sacramentalism that would better respond to the great needs of the world today. This is a missed opportunity for millions of the world's Christians. It is one of the most cherished ways many Christians believe they encounter God, and it serves as a weekly reminder for many that God can be met in and through common elements around a common table. But, instead of drawing their attention to God's all-pervading presence in the natural world, for many folks in the pews, the communion table instead reinforces the mediating power of the institutional church, its ecclesiastic jurisdiction, priestly privilege, and magisterial authority in the act of consecration. A most unfortunate preoccupation with the way bread and wine can be shared and with whom has more often than not eclipsed the earthiest aspects of sacramental theology.[57] It obscures the deeper significance of the communion table, that God can be known in and through the earthy things of this world. While some reconstruction has been done on this front, much more is necessary.[58]

---

55. For various social justice-based examples of how this question is asked and applied within a faith community's many contexts, see Elsbernd and Bieringer, *Theo-Ethic of Justice*, 187–213; also DeYoung, *Living Faith*; Massaro, *Living Justice*.

56. For a wide-ranging general overview of the cultural power of religion, from a cognitive-evolutionary perspective, and its ability to help the human mind deal with emotionally compelling problems, see Atran, *In Gods we Trust*.

57. Thomas Berry's suggestion that Christians place the Bible on the shelf for a while is partly his reaction to this concern that there has long been an over emphasis on God's expression through the written word rather than in the natural world. See Laszlo and Combs, *Thomas Berry*, 25.

58. My concern for reconstruction here is particular to the way Christian traditions might not only draw from and point to premodern creation-centered worldviews that work better with emerging ecological and evolutionary worldviews, but that there may be instances where new resources need to be created from within a living faith community. For the former, there are many creative efforts to do so. For example, see Delio et al., *Care*

The communion table can and should be imagined as a place where everyone is invited to engage each other, remember God's presence in the world, and reflect on the miracle of creation.[59] As a cherished tradition lying at the heart of religious life for so many people, the challenge is to shift focus from the limiting preoccupations with legalism and institutional authority to a new locus of ecological reflection.[60] During the Protestant Reformation, Martin Luther is said to have argued that God is as present in his cabbage soup as in the sacraments of bread and wine offered in the cathedral, albeit hidden in the former and revealed in the latter. His claim is sometimes interpreted as one that either downplays the importance of sacramentalism or denigrates how some believe they encounter God through bread and wine, but it is understood rightly as Luther's belief in God's all-pervading presence in the world.[61] This is the heart of sacramental theology and its central claim: God can be known and encountered in and through the things of this world.

Certain rituals in Christian traditions, such as the celebration of Holy Communion, can be particularly special reminders of God's all-pervading presence and activity in the world. An Augustinian framework of sacramental theology can help Christians today to reframe the communion table as a place that highlights earthy sacramentalism. William T. Cavanaugh argues that the consumption of bread and wine can be imagined two ways: Christians can consume the bread and wine by taking those elements into themselves, but they themselves can be thought of as consumed into

for Creation. Also, Ruether offers a helpful overview of the importance of the sacramental tradition within Christian traditions, though she cautions that the helpfulness of the sacramental traditions can only be responsibly recovered if they are reshaped in ways that free them of their patriarchal heritage. See Ruether, *Gaia & God*, 9, 229–53. These challenges and opportunities, of course, are certainly not unique to Christian traditions any more than sacramental theology might be considered so. Other religious traditions have their own rich history and practice of recognizing the sacredness of things like a grove of trees, a body of water, or a mighty river in arguably similar ways many Christians revere the sacred elements of the bread and wine or baptismal waters. Max Oelschlaeger argues that most of the world religions have within them much to draw upon and room to grow their general concern for Creation. See Oelschlaeger, *Caring for Creation*, 105.

59. This is a reference to Augustine's declaration that a single grain of any seed can inspire awe because it reflects the miraculous works of God (paraphrased). Margaret Miles cites this directly and adds that for Augustine, it is only a failure to order affections rightly and to use our senses that prevents us from experiencing this miracle of creation. See Miles, *Augustine on the Body*, 38.

60. See Galbraith, "Broken Bodies of God," 283–304.

61. Rasmussen, "Lutheran Sacramental Imagination."

something larger than themselves when they partake in the meal.⁶² Cavanaugh points to Augustine's belief in having heard God say to him "I am the food of the fully grown; grow and you will feed on me. And you will not change me into you like the food your flesh eats, but *you will be changed into me* [emphasis added]."⁶³ This Augustinian image of the self as drawn into the community via Christ at the communion table, as Cavanaugh notes, de-centers the individual or 'selfish' self by reorienting it in terms of the larger "context of a much wider community of participation [. . .]."⁶⁴ If that community of participation rightly includes the poor, disenfranchised, and most vulnerable, then the communion table is a place where their concerns become the community's concern.

The communion table is where a green and preferential option for the poor and oppressed is embodied. It is not enough for Christians to take a wafer and contemplate Christ's suffering on the cross while disregarding those whom Christ cared so deeply about during his life and ministry. It is where one meets God; it is there in the face of one's neighbor and the earthy things of this world. For those who partake in the meal from this perspective, it means the "very distinction between what is mine and what is yours breaks down [. . .] your pain is my pain, and my stuff is available to be communicated to you in your need."⁶⁵ This is not a touchy-feely, sunshine and butterflies kind of theology. This is and always has been a radical theology.⁶⁶ It is a call to ongoing transformation—a call to a process of continuous conversion of heart and mind toward the revolutionary mes-

---

62. Cavanaugh, *Being Consumed*, 54–55.

63. Cavanaugh, *Being Consumed*, 54–55. As noted, see Augustine, *Confessions*, book 7, chapter 16.

64. Cavanaugh, *Being Consumed*, 54–55.

65. Cavanaugh, *Being Consumed*, 56.

66. McFague provocatively asserts that "[u]nlike [. . .] first-century Mediterranean counterparts, North American middle-class Christians are not terrified by the unclean, but [. . .] are terrified by the poor." She argues that such fear is understandable if the sacramental nature of the Christian communion table is appropriately internalized, since it demands Christians acknowledge the sacred nature of the Earth and "share the planet's resources justly and sustainably with all" of the billions of poor around the world. She notes that the economic and political aspect of communitarian claims like these "demands basic changes in our economic policies toward greater egalitarianism at all levels." Moreover, consideration for the needs of the poor cannot remain limited to just those people who can barely subsist but should also be extended in some significant way to those species and ecological communities with whom all people share this sacred Earth. See McFague, *New Climate for Theology*, 93.

sage of the Gospel. It is a call to recreate one's self-image in terms of a larger community of being, including the larger family of creation. It is learning to replace a concern for personal salvation with a concern for planetary salvation. If Christian teachings and rituals are responsibly applied to the contemporary context, then the suffering of others *and* the degradation of life on this planet rightly occupies a central focus of Christian contemplation. This is earthy sacramentalism.

ENFLESHED SPIRITUALITY

While earthy sacramentalism promises to connect Christians to each other and the earth by way of the communion table, at the center of that celebration is a fully enfleshed spirituality. Enfleshed spirituality abides in an image of God in which God's transformative love for the world is represented by God's embodiment in the world. It is a form of spirituality emerging from a body theology rooted in the dirty, sweaty, fleshy experiences of human life. For Christians to connect the experiences of their lives and the stories of their faith to the larger story of planetary life on Earth and the creation of the cosmos, it is necessary to untangle the most helpful images of enfleshed spirituality from those aspects of personalist, salvation theologies that have come to dominate many streams of thought in the West during the modern era. Christian accounts of what it means to be human, of who God is, and how God acts in the world, need to change. Overly individualistic images of the human person and dualistic, disembodied images of God and world continue to feature prominently in Christian theology. During the last several centuries, those images have emerged in two notable forms: the image of a human mind and spirit as separate from the human body and from other people, and of God as above and apart from the world.

The first image presents an overly individualistic idea of what it means to be human before God. Sallie McFague describes this view of the self as too heavily emphasizing the desires and end goals of individual concern separate from the health and flourishing of the larger ecological and human communities to which all people belong. It was during the last three hundred years, as McFague argues, that Western societies have begun "internalizing an anthropology of radical individualism [. . .]."[67] From this perspective, there is an unfortunate tendency to sometimes objectify others,

---

67. McFague, *New Climate for Theology*, 48.

especially other forms of life. It is an image of the self as a self-regarding, highly rational being, oriented to the pursuit of individual freedom.

Postmodernism and liberation theologies, as McFague points out, critique this view of the self quite harshly as a "separate, individualistic, selfish, pretentious self who refuses to acknowledge its radical relativity."[68] McFague argues for an alternative view as one that is thoroughly embedded within a larger framework of creation, and therefore accountable to that larger community.[69] "The view of the self or subject that [could emerge]," as McFague describes elsewhere, "is not the individual who is 'saved' for life in another world, but a thoroughly embodied, relational subject who understands herself or himself as interdependent with everyone and everything else."[70] It is simply no longer appropriate, or accurate within the context of the world as we now know it, to imagine the self in those ways that came to dominate major streams of thought in the nineteenth and twentieth centuries.[71] Reimagining the self as an "ecological self," as McFague calls it, is a "functional activity" whose chief goal can be to "help the world prosper."[72] To see the human person as an embodied, ecologically embedded earth creature is to recreate an image of humanity as one whose flourishing is directly tied up with all other forms of life.

To imagine ourselves as ecological beings, need not diminish those special attributes of our species that make us different from other species. Various species on Earth certainly are not all the same by way of our shared characteristics, but we are all connected through a shared evolutionary story. As McFague acknowledges, "human beings are, at present,

---

68. McFague, *New Climate for Theology*, 48.

69. This is different from the way that some might describe the human person in terms of evolutionary biology. For McFague, a sense of the "self" remains, but others might interpret the ecological continuity of the human person so materially as to conclude that autonomous human consciousness, and regard for the "self," is wholly illusory. See, for example, Fromm, *Nature of being Human*, 245–46.

70. McFague, *Life Abundant*, 31.

71. While some thought forms of Christian theology have indeed stressed the order of creation and the work of divine providence for the vast majority of Christian history up to the modern period, generally they have been eclipsed by a turn toward the individualistic self and an impoverished view of God's activity in the world. A new view of the self, as an ecological self, and a view of God as embodied in the world could work together to inspire new ways for Christians to interact with each other and the Earth. Regarding Christian themes stressing the order of creation, see especially Johnson, "Losing and Finding Creation"; and Schaefer, *Theological Foundations*.

72. McFague, *Life Abundant*, 31–32.

the most complex developed creatures on earth," in terms of self-conscious awareness, even while we "share a common origin with everything else" and "are by no means the only distinctive creatures."[73] The evolutionary process birthed into our species the very special ability for self-conscious awareness; it is an ability that has emerged quite impressively in the human species even if it exists to some degree in other species on the planet.[74] The various distinctions between species, however, do not, or ought not, generate an arbitrary dividing line in some great chain of being that gives to one species the conceptual freedom to treat the Earth and all others in it as a kind of blank check to do with whatever the wealthiest and most powerful among us might wish.[75] Rather than sever our species from the rest of creation, this special distinction can lead us to see that, as McFague argues, "given our present numbers and power, we have the ability to be either for or against the rest of nature."[76] Said differently, "[w]e are not the only ones who matter, but we are the ones who are increasingly responsible for the others in creation."[77] As McFague frames it, humanity's evolved sense of self-conscious awareness includes a moral responsibility to recognize as inherently wrong the way we are collectively annihilating life on this planet. It is within our collective ability to change course. It is a moral choice the human species appears uniquely equipped to make for all life on Earth.

The second image pertains to Christian conceptions of God and of God's activity in the world. This is where McFague offers a particularly helpful account of God to counter overly dualistic images of God as a wholly transcendent being, unconcerned with the materiality of the cosmos. She paints a picture of the divine as metaphorically enmeshed within the earth's

---

73. McFague, *Life Abundant*, 46–48.

74. For examples of the robust nature of the discussion on this topic, see Terrace and Metcalfe, eds., *Missing Link in Cognition*; Lurz, ed. *Philosophy of Animal Minds*. See also Pinches and McDaniel, eds., *Good News for Animals*; Midgley, *Animals*, 134–43.

75. My reference here is to a hierarchical worldview, mentioned previously, that has persisted in Western philosophical and theological thinking from Plato through to the eighteenth century. See, for example, Lovejoy, *Great Chain of Being*. See also French, "Grace is Everywhere," 147–72; French, "Natural Law," 12–36. While French notes that Aquinas's emphasis on hierarchy is often poorly regarded as an ecological model in which lower creatures serve higher creatures in the "Great Chain of Being," he argues Aquinas's claim that each creature exists for the good of the whole order nonetheless reflects a higher good. See French, "Beast-Machines," 24–43.

76. McFague, *New Climate for Theology*, 47.

77. McFague, *New Climate for Theology*, 47.

ecosystems as its "source of life and vitality."[78] The technical term for this a panentheistic view of God and world; this is one in which God's presence is believed to be so utterly pervasive that God is perceived as dwelling both *in* and *beyond* all things simultaneously. Transcendent qualities of God have long been an enduring part of Christian traditions, and this view maintains them while at the same time highlighting the immanent qualities of God that have also long been part of the Christian moral imagination.[79] In McFague's image of the universe as the body of God, she imagines the world as an inspirited body in which the spirit of God "enlivens" and "energizes" the universe in the same way a team's "spirit" energizes it to win a game or in the way a spirit of resolution and vitality binds a group of people together, such as "in a common cause to oppose oppression."[80] God

---

78. McFague, *Body of God*, 145.

79. Today's contemporary context presents a challenge to Christian communities, inviting them to deemphasize those images of the self and God that have come to dominate streams of Western thought in the modern era. Christians can do this through a critical recovery of premodern creation-centered worldviews, grounded by earthy sacramentalism, and inspired by enfleshed spirituality. If God is imagined as the Creator dwelling in and amidst all things as source of life and sustaining vitality, then reconstructing Christian faith to better respond to the contemporary context requires a shift from human-centric, salvation-focused views of the world to those of an earthy nature, concerned with the flourishing of all life everywhere. My argument here is not against soteriology specifically, but rather its relatively modern predominance in Christian theology at the expense of creation themes. The "greening" of Christian soteriology may indeed be a fruitful avenue in addition to a renewed focus on creation as a subject itself. For various examples of ways theologians and ethicists endeavor to ecologize Christian soteriology, see the contributions in Conradie and Jenkins, eds., *Ecology and Christian Soteriology*, 107–265. As McFague puts it, it is no longer enough for Christians to simply focus on images of God concerning "me and my salvation nor even the salvation of human beings," but rather they must include concern for "the planet's well-being." See McFague, *Life Abundant*, 30. This rubs against overly transcendent and eschatologically salvific themes in Christian thought by drawing on those views of God's life-creating and life-sustaining work in the world as well as those views of the human as partner in that work.

80. McFague, *Body of God*, 143. The terms "spirit" and "body" figure prominently in McFague's metaphor and may initially appear to represent a heavily dualistic metaphor much like the mind/body metaphor, which asserts an understanding of God as orderer and controller of the universe. McFague's use of the term "spirit," however, is not as esoteric and dis-embodied as it may sound. See 144–45. Also, she refers to "spirit" in the same wide-ranging way it is used in common discourse and describes it as "a term with many meanings built upon its physical base as the breath of life." See 143. This is particularly interesting, given the earlier overview of those words in the Genesis accounts. In short, for McFague, "spirit" is not a term she intends to convey any kind of dualism; rather, she uses it in keeping with an integrated theology of the body. Her assertion that the world can be imagined as a divinely inspirited body is an attempt to create new views

is enfleshed in the cosmos in much the same way many Christians believe Christ is enfleshed in bread and wine.

Enfleshing Christian spirituality is about acknowledging the primary importance of body theology as an approach to God. It recognizes the importance of the human body and the human person's embodied experiences of the world in mediating one's encounters with God. It is also the starting point of Christian theological reflection. As James E. Nelson describes so eloquently, body theology starts "with the fleshy experience of life—with our hungers and our passions, our bodily aliveness and deadness, with the smell of coffee, with the homeless and the hungry we see on our streets, with the warm touch of a friend, with bodies violated and torn apart in war, with the scent of honeysuckle or the soft sting of autumn air on the cheek, with bodies tortured and raped, with the bodyself making love with the beloved [. . .]."[81] This right here, in these fleshy experiences of life, is where Christians start doing a theology that really matters.

We are, each of us, an enfleshed confluence of energy, emergent and emerging from a communion of multiple forms. We are the cosmos writ small. This makes it necessary to reposition focus away from things perceived as beyond the world and root it squarely in our experience of, and engagement with the world.[82] When Christians contemplate God and the world from the perspective of an enfleshed spirituality, they may come to see that, as McFague puts it, "our planet is a deteriorating body in desperate need" much like Christ on the road to Calvary.[83] Central to the gospel message is a call to respond to that great need of the world. So, the story of Jesus on the cross dying *for* human salvation is reframed as a story of a cosmic Christ who entered *into* the world as an embodied being—as an incarnate Creator dwelling in and amidst and in solidarity with its creation—simply

---

of God and world that both speak to contemporary crises and also offer up opportunities to become rooted in the traditions of the church.

81. Nelson, *Body Theology*, 42–43.

82. Elizabeth Johnson observes and then argues that "[a]ll contextual, liberation, feminist, and post-colonial theologies proceed with the realization that while dominant theologies may include 'the other' in some beneficial manner, the center of their intellectual and ethical interest remains the advantaged group, which does less than justice to those on the margins. The focus has to shift to those who have been silenced, so that their voices are heard and they are seen as of central importance in themselves. In a similar manner, the nascent field of ecological theology asks that we give careful consideration to the natural world *in its own right* as an irreplaceable element in the theological project." See Johnson, *Ask the Beasts*, xv.

83. McFague, *New Climate for Theology*, 94.

because God so loved the world.[84] There is no salvation from the world, only salvation with the world.

STORIES THAT MATTER

It is increasingly clear that for humanity, and especially the poorest and most vulnerable among us, the ability to survive and thrive is increasingly inseparable from the flourishing of the planet as a whole. Pounding people over the head with data does not help most people to see this or motivate enough to act. A new way of opening hearts and minds is required. Fr. Thomas Berry, a Catholic priest and a cultural historian of the world's religions by training, often highlighted the power and promise in stories for this task.[85] Berry saw ecology, evolution, and creation as essential resources for helping people shape the kind of stories into which they could imagine their lives.[86] I too think these sciences are a vital, and yet often missing, part of the new stories the world's religions need today.

Shaping and reshaping those stories people use to orient their lives in relation to the world is part of what Berry considered the great work of the present time. Early in Berry's academic career, and probably inspired in part by his direct experience of living in China in 1948, he focused on the

---

84. This is a reference to John 3:16.

85. Fr. Thomas Berry (1914–2009), named William Nathan at birth, was a Catholic priest who entered the Passionist Order in high school and assumed his new name because of his high regard for St. Thomas Aquinas. See Tucker, "Biography of Thomas Berry." In preparing this section and in reviewing the work of Thomas Berry, I am indebted to Mary Evelyn Tucker, John Grim, and the Forum on Religion and Ecology at Yale Divinity School for hosting "Living Cosmology: Christian Responses to Journey of the Universe" from November 7 to 9, 2014 in New Haven, CT. The conference yielded many excellent papers on Berry's work and I am particularly grateful for the personal stories and firsthand accounts of Thomas Berry offered by those who knew him well, including Tom and Catherine Keevey whose correspondence and gracious resource sharing has been invaluable to my understanding of Thomas Berry, the man behind his books. See also Keevey, "Thomas Berry, C.P."

86. Berry trained as a cultural historian, writing his dissertation on Giambattista Vico—a philosophical historian whose own accomplishment was to articulate operative patterns and delineate significant periods of time from history within a "big picture" context. See Berry, " Historical Theory of Giambattista." Vico's periodization of history is reflected in Berry's framework and naming of historical periods, which features quite prominently in his work, along with the seismic shifts of cultural awareness and cultural change that he argues happen during those transitions from one period to another. See Grim, "Thomas Berry's Vision."

history of cultures and religions in East Asia.[87] As President of the American Teilhard Association from 1975 to 1987, Berry began to develop his ideas around ecology and the story of evolution into what he coined the "new creation story."[88] This innovative intellectual project, or the creation of a unitive vision for human civilizations and a kind of grand narrative with implications for daily living, is one Berry saw as a necessary part of any way forward through what he was beginning to see as an ecological crisis. Grand visions are necessary to guide big changes, and he argued that the world needs stories big enough to "sustain human civilization in its transformation" from the present modern era of ecological destruction into a new ecological era or "Ecozoic Era" of responsible human-Earth relations.[89] Berry calls for a kind of myth-making on an epic scale, which is to say, one that is up to the equally epic task of ushering in "a period when humans would be present to the planet in a mutually beneficial manner."[90]

New myths are needed that are capable of penetrating the deepest levels of human consciousness. They are needed to inspire and motivate changes to deeply held assumptions one holds about oneself in relation to others and the world. These are assumptions ingrained in the larger frameworks of modern civilization, which Berry saw as fundamentally clouding human judgment at the most important moment in the ongoing development of human civilizations.[91] The Kantian notion of the world, for example, is one in which human reason is held as the sole source of subjectivity—one reducing other beings and the planet as a whole to objects of human study that may then be used, abused, and manipulated by human beings to whatever end they may deem necessary.[92] While Berry generally embraces modern sciences, he also surmised whether they sometimes might be prone to uncritically harbor this Kantian philosophical bias.[93] He

---

87. See, for example, Berry, *Buddhism*; also Berry, *Religions of India*.

88. See, for example, Berry, *New Story*.

89. Berry, *Great Work*, x.

90. Berry, *Great Work*, 3.

91. Berry, *Evening Thoughts*, 113-25.

92. For example, Kant posits that "[. . .] a human being really finds in himself a capacity by which he distinguishes himself from all other things, even from himself insofar as he is affected by objects, and that is reason." See Kant, *Metaphysics of Morals*, 57.

93. This concern sobers his otherwise positive embrace of the sciences, as seen in his statement, when he observes: "[t]he difficulty [. . .] with the rise of the modern sciences [is] we began to think of the universe as a collection of objects rather than as a communion of subjects." See Berry, *Great Work*, 16.

believed that if the human person is not adequately situated within an ecological context in which everything we do is measured against a standard of shared flourishing, then we find ourselves "ethically destitute" at a time when we could be transforming wisdom traditions at one of the most important crossroads in human and planetary history.[94]

Thomas Berry, along with physicist Brian Swimme, attempted the first telling of evolution as an epic story, which they called the new "Universe Story" in a book they published by that name.[95] In it, Berry and Swimme imagine an epic creation story informed by modern science but inspired and framed by a strong moral imagination and the kind of wonder and creativity often found in religious narratives. They tell the story of creation, as we know it through science, but in a way that opens up prevailing images of the human person to questions regarding our role in the cosmos, and our relationships with other forms of life. The story tells of cosmic, planetary, and human evolution in a way that emphasizes the integrated nature of those trajectories. They tell the human story as part of a larger story of life on Earth. As the story advances, the reader learns to see more clearly that humanity has an increasingly critical role to play—one where we are to collectively and actively work toward the preservation and conservation of all life in its various and wondrous forms. Human continuity and participation in the larger universe story frame perceptions of the human person with special or distinctive attributes as simultaneously wound up with and connected via a shared evolutionary story to every other being with which our species shares the planet.[96] Our consciousness, our curiosity, and our sense of wonder and awe are gifts the universe has brought into being through us, and we can use them to advance the flourishing of life across the cosmos.

For Christians, participating in the creative work of stories like these might mean turning to the book of nature as a kind of revelatory text. Christians can shape and reshape the way they think they ought to rightly live in the world by drawing inspiration from the natural processes of a good Earth they believe God called into being. Once while reflecting on a meadow, Berry observed that "[w]hatever preserves and enhances this meadow in the natural cycles of its transformation is good; whatever

---

94. Berry, *Great Work*, 104.

95. See Swimme and Berry, *Universe Story*.

96. With regard to a theological concern for human continuity with other animals specifically, sometimes this is called "creaturely theology." See, for example, Moore and Kearns, *Divinanimality*; also Deane-Drummond and Clough, *Creaturely Theology*. See also Derrida and Mallet, *Animal*; and Haraway, *When Species Meet*.

opposes this meadow or negates it is not good."⁹⁷ As a priest who had spent a lifetime pondering the Scriptures and observing the natural world, he came to believe quite firmly that "what is good recognizes the rights of this meadow and the creek and the woodlands beyond to exist and flourish in their ever-renewing seasonal expression even while larger processes shape the bioregion in its sequence of transformation."⁹⁸ Berry came to these conclusions as a Christian, but he imagined them as having wider-reaching effects across all spheres of human activity, not simply religion and philosophy, but also law and politics, economics and education.

The stories people rely on to inform who they think they are and the kind of world they think they are living in, really matter. They are shaped over time. They evolve sometimes quite slowly, but sometimes rapidly in response to a kind of earthquake of ideas in the prevailing popular consciousness. We are presently, as Berry so eloquently puts it, "between stories," wherein the "old story" is no longer adequate and a new story is needed.⁹⁹ Many of the old stories simply do not incorporate images of the world as we now know it, and they do not cultivate the kind of moral concern for other species, and the planetary life support systems on which we all depend. Critically and creatively imagining new stories is an essential way to usher

---

97. Berry, *Great Work*, 13.

98. Berry, *Great Work*. The kind of rights-based language he uses here is strong. To be clear, however, Berry is not calling for an extension of human rights to, say, a mosquito or to the meadows and creeks that inspired his ethic. Rather, he argues that "[t]rees have tree rights, insects have insect rights, rivers have river rights, mountains have mountain rights [. . . and notes that] [a]ll rights are limited and relative." See Berry, *Great Work*, 5. This concept also echoes a strong Thomistic framework, regarding the right use of property and the common good tradition, which Berry pulls forward. That connection is seen clearly in the claim Berry makes when he argues that people may "own property in accord with the well-being of the property and for the benefit of the larger community as well as ourselves." Berry, *Great Work*, 5.

99. Berry, *Dream of the Earth*, 123. Berry's writing draws inspiration from what he saw as creation-centered or "ecocentric" worldviews in some indigenous communities. He argued that some societies have been able to sustain themselves into the present age precisely because they operate out of an ecocentric worldview that acknowledges the limitations and sacredness of ecological systems. See especially his 1975 essay on "Historical and Contemporary Spirituality" in Berry, *Sacred Universe*, 49–65. Berry's vision includes a retrieval of ecocentric aspects of some premodern worldviews that he saw as having potential to inform those streams of thought that have come to dominate since the advent of the modern period. The creative aim and critical retrieval of these premodern stories ought not be to simply replace modern narratives—this is not about turning back the clock—it is about how they might inform, inspire, and shape entirely new stories going forward.

in a dramatic shift in the predominating collective consciousness. It can be an effort to help the world shift from a generally and overly anthropocentric approach to a responsibly retrieved and yet newly constructed creation-centered framework more capable of supporting planetary flourishing.

The moment at hand may very well be one in which our species initiates the dawning of a new era. For this to happen, human civilizations require a major paradigm shift in the way many people imagine who they are, and who they are in relation to others and the world. For Christians, this includes a shift in the way many see themselves in relation to God and the world they believe God loves. In a world grappling with the consequences of the modern turn, we now require yet another turn in the great wheel of life that births into the world new forms of collective consciousness about what it means to live a fully human life.

We have already explored the way some Christian traditions can be retrieved and reconstructed to facilitate this common challenge confronting us. In grappling with Scriptures and theological ideas spanning more than 2,000 years of Christian and pre-Christian history, I have pointed to several images I hope might awaken and sustain the hearts and minds of Christians today. In Genesis are images of a God that breathes life into earthen clay forms, where people are made, loved, and invited into God's co-creative love affair with the Earth. In Augustine, we find theological affirmation of Scripture's image of a good Earth. Moreover, people of faith are invited to imagine God as a vivifying power animating, sustaining, and directing all things. Christians are called to orient themselves toward that good. In Aquinas, we see a call to solidarity among those striving for the common good, and a God who binds human beings to each other through a shared responsibility for the common good. In the modern era, we find images of earthy sacramentalism and enfleshed spirituality that ground Christian faith in the world and tie up spiritual salvation with planetary salvation. We are invited to imagine the changes before us as epochal in nature. God is encountered in the stories we tell ourselves about who we are in the world, and so telling those stories well, is *doing* theology.

Questions remain: How do we move toward the kind of inspired action that can calm the screeches of darkness in our present moment and usher in a birdsong of planetary flourishing? How do human civilizations, and people of Christian faith in particular—those people in the pews and at the heart of contemporary Christian religious life whose daily actions wield such great potential for literally changing the current trajectory of human-Earth relations—how do they go about the hard, sweaty labor of physically reimagining their humanity in a world as we now know it? These are the questions and this is the turning point upon which the future of life on Earth hinges.

4

Into the Darkness with Hearts Ablaze

THE ENDANGERED SPECIES ACT of 1973 was passed by the United States Congress and signed into law by President Nixon. It established important legal protections for threatened and endangered species and their habitats, regarded as holding "esthetic, ecological, education, recreational, and scientific value [. . .]."[1] While that once groundbreaking legislation continues to afford important protections to threatened or endangered species, numerous efforts have emerged since then to extend legal protections beyond this baseline.[2] An example is the Great Ape Project. It was begun in the early 1990s as a scientific and moral call for the creation of a United Nations Declaration of Rights for Great Apes. It represents one of the first sets of next steps in expanded legal protections for other species.[3]

A number of countries are now introducing laws or even constitutional provisions extending protections to other species. Switzerland was

---

1. The Endangered Species Act (ESA) of 1973's predecessor legislation includes the Endangered Species Preservation Act of 1966. The ESA is administered by the United States Fish and Wildlife Service (FWS) as well as the National Oceanic and Atmospheric Administration (NOAA). See US Fish & Wildlife Service, "ESA Basics"; and Stanford Environmental Law Society, ed., *Endangered Species Act*.

2. There are several examples of the ongoing effort to create legal space for the needs of other species by extending animal protection laws beyond those that categorize other species according to their human use. For some brief examples of the literature, see: Eisen, "Liberating Animal Law." See also Linzey, *Animal Theology*; Linzey and Cohn-Sherbok, *After Noah*; Linzey and Yamamoto, *Animals on the Agenda*; Linzey, *Why Animal Suffering Matters*.

3. Cavalieri and Singer, *Great Ape Project*.

one of the first to pass a constitutional amendment in 1992 recognizing animals broadly as "beings" rather than as "things," and Germany later added a clause to its constitution obligating the state to expand its respect and protection for human dignity to include respect and protection for "animal dignity."[4] Spain's parliament passed a toothy resolution in 2008 that banned medical experimentation on, and made it a crime to kill, any member species of the great apes, which includes gorillas, chimpanzees, bonobos, and orangutans.[5] Recognizing the inherent rights of other animal species, and safeguarding them by extending necessary protections, reflects a monumental evolution of modern governance in the West.[6]

Moral concern for rights held by nonhuman beings is not limited only to other animal species. The concept has been extended to include more broadly construed rights for nature. Such broader-based notions of rights recognition may include whole ecosystems or even bioregions. Some regard rights for nature as a continuation of the modern environmental movement beginning in the United States during the 1960's and 1970's and then spreading around the world. Others contend, however, that the roots of a movement conceptualizing rights for nature and ecological systems are actually quite ancient. Still, others question whether a rights-based approach is even capable of leading to the kind of preservation or conservation natural systems increasingly require in a world where resources seem so constrained.[7] Perhaps such an approach, rather than an end goal of environmentalism, simply reflects the fits and starts of a world only just beginning to conceptualize a response to a planet groaning under the weight of modernity's heavy yoke.

It has been mostly indigenous communities taking the lead in this area, with premodern religious ties to landscapes, bodies of water, mountains, and the like that they have long considered sacred. When the Māori

4. Associated Press, "Germany Guarantees Animal Rights."

5. Cohen, "What's Next in the Law"; Abend, "Human Rights for Apes."

6. The ongoing debate over nature rights has developed into an expanded discourse in philosophical circles with varying ideas on the appropriateness of a rights-based approach with regard to animals and nature. For a sampling, see Povilitis, "On Assigning Rights," 67–71. See also Watson, "Self-Consciousness and Rights," 99–129; Norton, "Environmental Ethics," 17–36; Hargrove, *Animal Rights*; Klyza, "Do Trees have Rights?," 427–44; Waks, "Environmental Claims," 133–48; Singer, *Animal Liberation*; and Regan, *Case for Animal Rights*.

7. For some of the pitfalls of a rights-based approach versus a value-based approach, see Rolston III, *Conserving Natural Value*, 17–18. On navigating conflicts, see Rolston III, "Feeding People versus Saving Nature?"

tribe won recognition for the Whanganui River to be accorded its own legal identity under New Zealand law, the river was granted the rights, duties, and liabilities of a legal person.[8] On the heels of that decision, a north Indian court in the state of Uttarakhand accorded the Ganges and Yamuna rivers the status of a legal person "with all corresponding rights, duties and liabilities" under Indian law, citing the precedent set by New Zealand.[9] It appears as though the world's governing bodies are responding to a new but inherently ancient call to action. There may be better ways for governments to respond to those persistent voices that dare to rise up and demand a different way of doing things.

Whether the roots of this phenomenon are ancient or post-modern, they reflect a significant expansion in the way moral worth and legal standing are accorded within the emerging collective consciousness and sociopolitical moral imagination of the world's governing bodies. But, this is only one way forward—there are others. We will explore the development of this dynamic movement in the next section.[10] We will also look at the way changes to market systems, and the way economic value is accorded, are helping to foment a paradigm shift in the way people live on a shared planet with finite resources. Then, we turn to the call for a theology of mobilization that bonds together faith and action with a vision of social justice and ecological responsibility. These examples of progress offer what I hope is the promise of a bright morning, only now just beginning to dawn over the horizon. It is, I hope, a promise capable of inspiring us all to work with bold faith and moral courage in the world as it is, for a world as it could be.

EXPANDING THE MORAL IMAGINATION

The question of nature's legal standing under US law has been debated before the country's highest court. In 1972, the Sierra Club argued a case against permitted development in Mineral King, a glacial valley near Sequoia National Park in California. The Club lost because it was deemed to lack standing. Specifically, the court ruled that the Club could sue only

---

8. Roy, "New Zealand River."

9. Safi, "Ganges and Yamuna Rivers."

10. Whether it is morally or even legally helpful to regard nature as imbued with inherent rights to be recognized by political establishments is a complex, long-debated question. For examples offering an overview of its early development, see Nash, *Rights of Nature*; Stone, *Should Trees Have Standing?*

on behalf of its members. The Court held that only local residents would have legal standing to bring the case forward, if they could contend that the development would cause them tangible harm. No such residents came forward, and the Court ruled that the Sierra Club could not bring the case forward on the claim of representing the interests of the ecosystems alone.

The Sierra Club lost the case, but the ruling was heralded as a victory for environmental organizations. It would facilitate the future legal challenges of people on behalf of environmental protection, though notably, it would not recognize standing in nature. The dissenting opinion of Justice William O. Douglas did, however, reason that "inanimate objects" should have their standing recognized in US courts. He argued:

> The critical question of "standing" would be simplified and also put neatly in focus if we fashioned a federal rule that allowed environmental issues to be litigated before federal agencies or federal courts in the name of the inanimate object about to be despoiled, defaced, or invaded by roads and bulldozers and where injury is the subject of public outrage. Contemporary public concern for protecting nature's ecological equilibrium should lead to the conferral of standing upon environmental objects to sue for their own preservation [...].[11]

The ruling decided that the Club could not sue on behalf of nature itself since nature and natural systems like rivers, lakes, and estuaries traditionally do not have standing. They could only sue if failure to protect the environment threatened human members who do having legal standing before the court. Since this case, the idea that ecosystems may be regarded as rights bearers with legally recognized standing has been put forward by some municipal governments in the United States, such as those in New Hampshire, Pennsylvania, and Virginia.[12]

Internationally, the idea has had more traction and nations in Latin America have been among some of the first to recognize nature's standing. In 2008, the Ecuadorian people voted to approve a new constitution that gives nature "the right to exist, persist, maintain and regenerate its vital cycles, structure, functions and its processes in evolution."[13] A few years later, the Bolivian government recognized the inherent rights of *Pachamama*, or

---

11. Sierra Club v. Morton, 405 U.S. 727, 741–43 (USSC 1972).
12. Revkin, "Ecuador Constitution."
13. Revkin, "Ecuador Constitution."

Mother Earth, as its own legal entity and Bolivia recognized nature's rights to exist, to continue vital cycles, and not to be polluted.[14]

Known by various other names to indigenous communities across Latin America, the word *Pachamama* conceptualizes an ancient sentiment that the divine animates or embodies all that a person sees around them—what others sometimes refer to simply as nature or perhaps more reverently as "Mother Earth."[15] *Pachamama* is originally an Andean term that has received more attention than other indigenous concepts for Mother Earth, because it has acted as something of a uniting force, bringing together otherwise disparate indigenous communities around a common political cause, rooted in a shared spiritual concern for *Pachamama*. The concept is the bedrock of a social and political movement that connects the need for legal protections by two traditionally vulnerable populations in Latin America: indigenous communities and the ecosystems they depend on and revere.[16] While those nations recognizing nature as *Pachamama* have been some of the first to grant it status under the law, only time will tell whether that recognition and the laws on the books will be consistently enforced. If the law holds, it is probably because it merely enshrines into law an idea already held dear by the people—the land is sacred and is worth preserving.

Affording nature some regard for its own intrinsic moral worth through formal legal recognition is inherently a modern advance, albeit in the Bolivian case it is one with roots in ancient indigenous conceptualizations of the Earth. The way it has played out in Latin America, and in New Zealand too, demonstrates something of an emerging hybrid of pre-modern and modern worldviews. Carolyn Merchant describes one pre-modern image of Earth as "a living organism and nurturing mother," that for much of human history "had served as a cultural constraint restricting the actions of human beings."[17]

With the advent of the Scientific Revolution, there were important changes to the way some premodern conceptualizations imaged Earth and so envisioned proper human action in relation to it.[18] Many behavioral and

---

14. Radicella, "Protecting Pachamama"; Vidal, "Bolivia Enshrines Rights."

15. Núñez and Larraín, *Patrimonio Arqueológico Chileno*. As cited in Castro and Aldunate, "Sacred Mountains."

16. Becker, "Indigenous Movements"; O'Connor, "'Part of Nature.'"

17. Merchant, *Death of Nature*, 3.

18. That is not to say that all premodern worldviews are alike or that they always resulted in more ecological and sustainable societies. Some of those worldviews, including indigenous worldviews, may be considered earth affirming and ecologically sustainable.

cultural constraints disappeared during and after the Scientific Revolution so swiftly and thoroughly that it is perhaps more fitting to say, as Merchant does, that those constraints actually morphed into "a sanction" for exploitation.[19] It may be that what is now occurring in some modern legal and political systems, is an effort to reclaim this ancient sense of human intimacy with the Earth and its natural systems but from within a modern social framework.

There are aspects of a modern shift, however, that ought to be preserved, if ancient ideas are reclaimed. The general movement toward the liberation and the expansion of rights for more and more people in modern governance is chief among them. So too is the scientific method as a way to learn about the world and ourselves. If these are preserved in a way that harmonizes with an ancient conceptualization of Earth as sacred, and if they can be pushed forward together as twin priorities for the flourishing of all life on Earth, then we may find ourselves on the precipice of one way forward. In creating what Merchant calls a "partnership ethic," communities are able to reclaim certain premodern images alongside selective modern frames.[20] Premodern ideas can resurface and penetrate prevailing worldviews in ways that powerfully disrupt modern and postmodern habits of denigration and degradation. Metaphors for the Earth as an "active partner" as Merchant calls it, can borrow from a rich pool of images developed by diverse peoples across the globe. Such metaphors have long been used by people to make meaning and to connect with forces they believe to be at play in the universe.[21] The challenge is to view this as a continuous

---

Some may not. For case specific examples of the latter, see Nayamweru, "Women and Sacred Groves," 41–56; and Lorentzen, "Indigenous Feet," 57–72.

19. Merchant, *Death of Nature*, 4. Also, Jerry Mander points to a shift from what he considers to be the inherent wisdom found in some indigenous "systems of logic" toward what he calls "American technological society," which he argues is rooted in a myth of progress through technological advancement that contributes to the kind of ecological violence and industrial expansion occurring in the modern era. See Mander, *Absence of the Sacred*, 39, 220.

20. Merchant, *Reinventing Eden*, 205–6.

21. For an excellent introduction to indigenous cosmologies, see Grim, ed., *Indigenous Traditions*. See also Grim, *Shaman*. A wide-ranging investigation into the many different perspectives on Nature, including indigenous perspectives, that have emerged from (and evolved) within the North American context, are noted in Albanese, *Nature Religion in America*. For contributions on the perspectives of North American Indians on this topic, see Hughes, *North American Indian Ecology*; Marshall, *To You we Shall Return*; and Lushwala, *Time of the Black Jaguar*.

process in which the moral imagination is frequently stretched in light of new knowledge and in response to new problems.

We need new metaphors to help paint a picture of what it means to lead a good life in relation to our neighbors and to the one world we all share. Whether these images emerge from ancient sources, it is critically important they ultimately help circumvent any historical legacies of oppression in all its forms. This will likely require more robust scaffolding than that offered by a rights-based framework alone. But, any way forward will have to be every bit as pragmatic as this approach tries to be. The principles of faith, equity, and ecology form a triumvirate upon which a strong foundation and a new paradigm for the future may be constructed, and it is one that is grounded in the everyday lives of ordinary people.

FOMENTING A PARADIGM SHIFT

The taproot of the contemporary ecological crisis is the overconsumption of resources by the world's most privileged peoples. The global appetite for consumer products, along with a growing human population, has reached epic proportions that are increasingly exacerbated by a metastasizing "throw-away" culture. The average US citizen alone is now consuming planetary resources more than five times beyond what the Earth can sustain.[22] Also, the production and consumption of energy and energy-intensive resources are one of the primary driving forces of climate change and environmental degradation, as well as the burdens this places on the planet's most marginalized communities. The word's relative elite is overconsuming even as majorities of the world's people on the lowest rungs of the economic ladder are not consuming enough to meet even their basic needs.

Work is afoot to factor both ecological value and social justice into economic calculations in order to help reshape prevailing market systems so they preserve planetary ecosystems and better sustain the world's most vulnerable people dependent on those systems. Some short-term solutions might go a long way in making the world a more just and humane place for a larger number of people today, while a deeper long-term shift in human-Earth relations will inevitably be necessary. Care for the Earth and care for

---

22. Many popular "ecological footprint" calculators use the language and calculations discussed in Boyd, "One Footprint." The soundness of such language and calculations as a responsible way to discuss sustainability, however, is not without critique. See Fiala, "Measuring Sustainability," 519–25.

the poor is the primary principle by which such a major paradigm shift ought to be structured to help course correct our present trajectory.[23] Both short-term and long-term solutions must address the disproportionate nature of resource consumption.

Overconsumption by the world's wealthiest populations is facilitated among the most developed countries partly because the costs of production and consumption are not fully accounted for within the dominating economic orders of these societies. Since private companies typically are not charged for spewing waste products from their production process into the global commons, including wastes such as carbon dioxide or soot produced during energy consumption, they need not worry about creating a surplus of those wastes.[24] When consumers flip on a light switch or kick up the heat, their energy bills do not reflect the environmental cost of fossil fuel emissions spewing into the atmosphere. Neither does the cost of fuel usually reflect the full price of its extraction and transportation, resulting in artificially "cheap" prices. So, when industries and the companies that profit from them are not held to account for these costs, then the costs of safely

---

23. Thomas S. Kuhn, a physicist and historian of science, introduced this term "paradigm shift" in 1962 as a way to describe periodic shifts or revolutions in the advancement of scientific knowledge, and particularly those shifts brought on by some sort of crisis within the *"gestalt"* of the scientific community. See Kuhn, *Structure of Scientific Revolutions*, 85–89. One example of a paradigm shift is the shift in consensus from the Geocentric Model that was part of the Ptolemaic system to a Heliocentric Model developed and exchanged for the Copernican system in the seventeenth century. For a general overview, see Kuhn, *Copernican Revolution*. Also, consider how the theory of plate tectonics and continental drift, as a more contemporary example, has advanced scientific understandings of evolutionary biology, ecology, and earth sciences. Such an advance was not only seismic in terms of how it helped to advance scientific understanding of planetary evolution and functions but demonstrates how paradigm shifts often emerges only out of the collective work of many individuals and groups over an extended period of time, sometimes generations. See, for example, Shubin, *Universe Within*, 84, 115–16, 201–2. Finally, while Kuhn limited his use of the term to describe shifts or revolutions in science, the comprehensive and sometimes highly disruptive nature of those shifts upon prevailing worldviews, as well as the lengthy time scale on which such shifts can take place, allowed his idea to transcend Kuhn's originally intended use for the term. I think it applies comfortably to a description of the kind of transformation presently underway. Humanity is on the cusp of a major paradigm shift because it is challenging entrenched aspects of prevailing modern worldviews. Consider how Edwards uses it with regard to "sustainability" as a social revolution on par with (and in response to) the Industrial Revolution, in Edwards, *Sustainability Revolution*, 1–10.

24. For more on the global commons, see Goldman, *Privatizing Nature*; Harrison and Sundstrom, *Global Commons*; Milun, *Political Uncommons*; Jasper, *Conflict and Cooperation*; and Amstutz, *International Ethics*.

processing waste, when it does happen, falls unfairly to the larger society as a whole. Safe processing of some wastes, however, is not always possible and not always undertaken. The atmosphere and the land absorb it and it accumulates in our bodies and especially the bodies of the poor.

Famed environmentalist Lester Brown once met with the former Vice President of Exxon for Norway and the North Sea, Øystein Dahle. Dahle conveyed to him his opinion that "[c]apitalism may collapse because it does not allow the market to tell the ecological truth."[25] Neglecting the ecological costs of production and consumption reflect not only a power imbalance arising from the disproportionate privileges of the global elite relative to everyone else, but it may be the single largest contributing factor to global climate change. At bottom, climate change is the result of the overproduction of carbon-heavy industries as well as the disproportionate overconsumption of its products. As a global phenomenon rooted in structural market inefficiencies, it disproportionately harms the poorest of the poor.

Jeffrey D. Sachs argues that the efficiencies of free markets "[f]ail when producers cause adverse spillovers to the rest of society, such as by polluting the rivers with toxic chemicals or emitting climate-changing carbon dioxide into the air from a coal-fired power plant."[26] He notes that in such cases, "the private economy tends to oversupply the goods in question, unless there are specific regulations or levies imposed on the offending actions."[27] The spillovers to which Sachs refers are also called externalities.[28] William C. French argues that "[b]y leaving these real but

25. Brown, *Plan B 4.0*, 243.
26. Sachs, *Price of Civilization*, 33.
27. Sachs, *Price of Civilization*, 33.
28. See also Gore, *An Inconvenient Truth*, 270; Brown, *Plan B 3.0*, 8, 290, 367. Though externalities are not factored into the economic costs of production and consumption, they carry enormous untallied social and environmental costs. When the costs and consequences of production and consumption, or all the inputs and outputs required by an accurate economic balance sheet are not accounted for, then short-changing occurs somewhere else. Responsible accounting is the foundation of robust market systems. Neglecting it portends market vulnerability, and risks eventual failures. This is particularly disconcerting for those around the world with the most to lose from the widespread failure of such systems. Whitney Bauman argues, for example, that consumerism or the promise of access to wealth and goods, can act as a "carrot" for poorer peoples to buy into capitalism as an economic system even though it may not be as effective at lifting as many people out of poverty as believed, at which point military and economic power concentrated in the hands of a wealthy elite then act as a "stick" to maintain a system that funnels wealth upward. See Bauman, "Consumerism and Capitalism," 263–64.

widely-diffused costs out of the price paid in the market transaction, we get an immense privileging of the present interests of some coupled with a structural blocking of necessary concern for burdens on others [...]."[29] In other words, "[s]hort-term concerns by some for profits and power are allowed to swamp long-range responsible planning for the common good."[30] This may be less a problem of greed, per say, and more a basic failure in the design structure of the economic system.

Design flaws can be corrected, but this is where ignorance, greed, self-preservation, and the application of privilege and power do come in and present a challenge. The environmental degradation and the unjust suffering experienced by economically disadvantaged communities living in polluted areas occurs while stockholders in large, multinational corporations like ExxonMobil tend to live quite far from the pollutive externalities generating their profits. This includes not only the wealthiest global elite but also many middle class people who have these shares in their retirement portfolios and are not even aware they are there. These companies and their shareholders profit while the communities in which they extract and produce their commodities generally pay the uncounted costs of production. Those living in such communities suffer in terms of reduced health and quality of life from the search, production, transport, and consumption of non-renewable resources like oil, coal, and natural gas. Most people who own shares own so little that divesting probably would not change much, and few incentives exist for those who benefit most from the system to do anything at all in the way of changing it.

Citizens and consumers can, however, take action collectively. We see this happening when students demand their university's multi-billion dollar portfolio divest from coal and oil or demand their governments hold corporate enterprises accountable. People can and indeed are already insisting on better accounting for the full costs of production. On the one hand, this is about values of basic fairness in which those who make a mess and profit from it are held responsible. It is patently unfair to ask the world's most vulnerable ecosystems and the most economically impoverished communities to bear the brunt of the extractive economy and its pollutive industries. On the other hand, requiring market systems to take a full account of costs can actually contribute to a stronger, more robust market system that in turn contributes to a larger common good by helping markets become generally

29. French, "On Knowing Oneself," 146.
30. French, "On Knowing Oneself," 146.

less vulnerable to debilitating, system-wide failures.[31] Citizens of the world's most powerful democracies can, and should, require their governments to update market systems so they better contribute to a world that works better for more people.

Market systems predominating in the West generally took root at a time and from a vantage point in which nature's bounty and ability to absorb human waste could have seemed nearly endless. But nature's bounty can no longer be rightly seen as infinite and this is a different challenge from the one posed by the problem of externalities. It reflects a very new change in circumstances that the human species is confronting as an increasingly technological, globalized society. Aside from the devastation wrought by natural disasters, which have a way of reminding people of nature's persisting power and agency, nature is generally and increasingly perceived as fractured and vulnerable. Rather than vast expanses of wilderness, the world's largest populations are increasingly urban with fewer opportunities to engage in meaningful interactions with nature beyond highly constrained environments. Phenomena such as climate change, ocean acidification and coral bleaching, melting glaciers, and poor air quality are markers of nature's growing inability to accommodate the status quo of business as usual. Market systems can and ought to evolve, in light of these new circumstances, into a force that positively contributes to the long-term transformation of human-Earth relations. Specifically, market systems of production and consumption need to become much more sensitive to the changing needs of societies, and increasingly that includes the need to function within the ecological boundaries of what planetary systems can sustain.[32]

---

31. We now live in a world where it is so overwhelmingly prudent to protect and conserve natural resources, that the financial benefits of preserving and protecting life outweigh even the financial costs of ignoring the externalities, especially when considered in the larger context of the global commons. A study backed by the United Nations and several countries has assembled a report on "The Economics of Ecosystems and Biodiversity" (TEEB) in order to help global society begin accounting for the economic value of the ecological services that ecosystems provide human societies. See Sukhdev and Brink, eds., *TEEB*. Also, Bishop, *Economics of Ecosystems*; Stern, *Economics of Climate Change*. Pavan Sukhdev of Deutsche Bank notes that all of the thousand-plus studies the group has evaluated regarding efforts to protect ecosystems around the globe point in the same direction; specifically, he says that "no matter how you slice the figures up you come up with a ratio of benefits to costs that's between 25-to-one and 100-to-one." See Black, "Profit from Nature Protection."

32. On the future of capitalist systems, see Rifkin, *Zero Marginal Cost*.

Better market systems can affect transformational change across sectors from housing, transportation, energy, and food systems all the way to how business is conducted. Where capable regulations are in place to encourage market efficiencies, we get a glimpse of a world that generally works better for more people and other forms of life. This includes major gains in energy efficiency for home appliances, lighting, building codes, and building management standards. Smart grids, expanded mass transportation infrastructure, as well as the design of walkable, bikeable, dense urban centers around the world—all are revitalizing cities and creating a broadly reduced need for fuel and energy consumption. Reduced subsidies for pollutive industries such as coal and oil in some markets have already incentivized the shift toward more renewable sources of energy such as wind and solar, and where appropriate also biomass, hydropower, and geothermal; occasionally, entire countries have used policy changes to re-center their economies around renewables, as in the case of Germany. This redirects economic growth toward industries that are gentler on the earth and cleaner for the people who live and work in the communities nearest them.

Highly subsidized industrial food systems are being positively transformed too, albeit slowly, toward more efficient, productive, sustainable, and accessible forms of agriculture, via the local, organic foods movement, new urban farming collectives, community supported agriculture farms, and food cooperatives. In the United States, small family farmers struggle to grow fresh fruits and vegetables and are forced to compete at market against highly subsidized agri-industries selling highly processed food products, which are often composed primarily of heavily subsidized agri-products like sugar and corn by-products. These foods are generally not as good for most of us as say, a carrot, and yet our governments continue subsidizing them instead of the family farmers who grow fresh produce on a sustainable farm. Why? We have a long way to go, but things are starting to change as more people choose local, fresh, responsibly grown foods. The fresh, organic food movement in the United States continues to expand and blossom despite many challenges. It shows the power of consumer-driven demand and the ability of consumers to transform a market sector with the choices they make at the grocery store.

Meanwhile, certified benefit corporations or "B Corps" are for-profit companies that voluntarily include social and environmental responsibility, accountability, and transparency as an integral part of their mission and business practices.³³ For B Corps, performance is judged by a triple bottom line that includes not only the company's financial health but also its positive social and environmental impacts too. Increasingly, even traditional companies are adding Corporate Social Responsibility (CSR) officers to their c-suites. They are instituting policies and championing causes prioritizing more than financial performance alone and changing the view of shareholders as sole stakeholders. Initiatives promoting workplace diversity and sensitivity, or that consider environmental concerns such as the carbon footprint of a supply chain may have fewer direct bottom-line connections but offer positive contributions to society at large. They also do a better job of recognizing as stakeholders a company's employees, its customers, and the communities in which they operate.³⁴ Many of these changes are only in their infancy, and some amount to little more than grandstanding or greenwashing; but some efforts are the result of innovative business and industry leaders who have determined they will swim against the current of business as usual because it is the right thing to do. Imagine how much more could be possible if we were to change the flow of the current itself. This is possible—we have done it before. We can, and should, do it now.

The historical evolution and development of social systems warrant a healthy dose of optimism for the possibility of future change in light of current challenges. The stories that orient human lives and the institutions and systems people create have on occasion grown into more equitable, more just, and more sustainable frameworks. I am optimistic humanity can collectively navigate through the thorny, uncharted territory before us if we orient ourselves with informed, thoughtful visions of ecological and social justice. While there are likely many more ways to do this, some of the examples to which I have pointed are already moving us much closer to meeting these goals. They likely may not, however, take us as far as we are capable.

---

33. See Honeyman, *B Corp Handbook*. Also, "What are B Corps?" *Certified B Corporations*.

34. Consider the significance of a letter written by Laurence D. Fink, chief executive of BlackRock—the largest investment firm in the world, managing more than $6 trillion in investments—to the chief executives of the world's largest public companies. In it, Fink essentially informs business leaders they will need to do more than make profits if they will want to receive BlackRock's support. A copy of his letter and summary of its significance can be found in reporting by Sorkin, "BlackRock's Message."

A THEOLOGY OF MOBILIZATION

Strong, mobilizing forces are animated by clear, galvanizing visions of what people believe the world could and should become. A vision of social justice rooted in ecological responsibility is partly one in which the globe's poorest people are empowered to claim their agency, equality, and access to basic resources. It is a world in which communities, institutions, and social structures from the local to the global level are increasingly ordered and restructured to facilitate the transformation of all forms of oppression and injustice into more just relations that enhance and preserve human dignity and ecological integrity. This is no small task; it is, as Thomas Berry names it, the great work of this generation.

Ecological responsibility in the context of this vision can be understood as a necessary component of social justice, even as the goals of social justice necessarily must include the pursuit of ecological integrity. William E. Gibson describes an ethic of eco-justice as one that "recognizes in other creatures and natural systems the claim to be respected and valued and taken into account in societal arrangements [. . .]."[35] Moreover, he suggests that "it sees humans as bound up with, and integral to, that larger living fabric of all that is, which some call simply nature and some call God's good creation."[36] An integrative ethic of eco-justice recognizes the deeply entwined concern for justice among human beings and also patterns of justice promoting the flourishing of all life on the planet.

For Christians, an ethic of eco-justice is really an expanded form of justice as participation. Justice as participation recognizes the relationality and interdependency of all life on our shared planet and correlates it to a long-standing tradition of Trinitarian theology as well as to an embodied, relational view of the human person.[37] For Mary Elsbernd, OSF and Re-

---

35. Gibson, *Eco-Justice*, 24. Also, Hessel, *After Nature's Revolt*; and Hessel and Rasmussen, *Earth Habitat*. See James Gustafson for a discussion of "relevant whole" as it relates to understanding the human person in terms of a larger ecological and cosmic context in Gustafson, *Theocentric Perspective*, 219–50.

36. Gibson, *Eco-Justice*, 24.

37. Elsbernd and Bieringer understand justice as participation, with subsequent quotes noted in full here, to be "coherent with those understandings of God which highlight God's universal invitation to enter into relationship with Godself and to continue the works of God, namely, creation, liberation (intervention), and resistance (judgment) to injustice. The image of God as a trinity of persons constituted in their difference provides additional theological underpinnings for justice as participation. Finally, our description for justice as participation is connected to an anthropology which recognizes

imund Bieringer, justice as participation draws on "those understandings of God [emphasizing] God's universal invitation to enter into relationship with Godself and to continue the works of God, namely, creation, liberation (intervention), and resistance (judgment) to injustice."[38] Elsbernd and Bieringer argue that justice as participation is rooted in an understanding of the human person that recognizes the constitutive characteristics of "embodiment, social location, relationality, fundamental equality in originality, and accountable agency."[39] It is God's universal invitation to all people to be in relationship. It is not only in relationship with the divine, but also in just participatory relationships with one another, and with the world in and around us that people are called to experience the fullness of their humanity.[40]

In terms of concrete, everyday, lived experiences, what does it mean to cultivate a life of justice as participation? It means different things for each of us, though nobody is excluded from the privilege and responsibility of participation. For those of us in the developed world where many of us have lifestyles marked by overconsumption, we are at least partly complicit in the global phenomena that must change. For us, our invitation may be to a life of greater simplicity, generosity, service, and hospitality to those with less power, less privilege, and fewer resources at their disposal. It almost certainly means consuming less when possible, and consuming more responsibly when and where we can. It may mean yielding the floor to those in the public square who have been historically underrepresented, or it may mean stepping up to claim your voice and your space in that place. Collectively, we all share responsibility in striving to create cultures of encounter, rather than confrontation in our schools, our workplaces, our businesses, our houses of worship, and civic spaces. We are each one of us called to participate in both the individual and collective, structural change-making

---

the following constitutive characteristics: embodiment, social location, relationality, fundamental equality in originality, and accountable agency." See Elsbernd and Bieringer, *Theo-Ethic of Justice*, 160.

38. Elsbernd and Bieringer, *Theo-Ethic of Justice*, 160.

39. Elsbernd and Bieringer, *Theo-Ethic of Justice*, 160.

40. As Holmes Rolston III astutely points out, reaching out and embracing nature in Christian love may involve an embrace that simply lets nature be. Especially with regard to the most vulnerable ecosystems, and to wild animals in particular, they "do not need our beneficence, they need us to leave space and let them alone." Rolston III, "Loving Nature," 315.

that is such an important part of a life well lived and of building our shared future.[41]

Justice as participation also means extending our time, talent, and resources to accompany others in the struggle for justice. This starts in our nearest circles but extends beyond them too. That may mean standing alongside a colleague who is paid less for her work because of her gender, or speaking up when someone on the bus is harassed because of how they look, whom they worship, or whom they love. Farther afield and to the degree we are able, we are called to stretch almost beyond where we can go, and consider how our time, talent, and resources might accompany the struggle for justice wherever it might pop up. How can we facilitate smart development policy and good governance in our community, in our nation, or around the world? How can we contribute to a global culture in which the most vulnerable and marginalized among us are able to participate in shaping the social structures and policies that govern their lives? How do we address major problems like climate change by incorporating the culturally differentiated needs of women and men at all levels of work?[42] How do we promote space for each person to claim a greater sense of his or her own agency, power, ability, skills, gifts, and participation in society and in the wider world? The aims of social equity and environmental sustainability are each more achievable when they are both treated together. Social equity and sustainable development go hand in hand; this is the structural embodiment of justice as participation lived out in the real world.

---

41. This is a reference to discussions on normativity of the future, in which justice can be visualized as the eschatological City of God breaking into the present. Said differently, when one prays, "Thy Kingdom come, Thy will be done on Earth as in Heaven," it is possible to imagine a hoped-for world of justice and peace that is not yet here and never experienced fully, but yet which peaks around the corner offering occasional glimpses of itself. Those glimpses help to inspire the good work of the Peaceable Kingdom as they burst into the present moment offering hope and inspiration from future possibilities. See, for example, Bieringer and Elsbernd, eds., *Normativity of the Future*.

42. For an example of how this has been done successfully in terms of international development policy, see the work of Monique Barbut, who was CEO of the Global Environment Facility, the world's largest funder of efforts to preserve the global environment, from 2006 to 2012. She has done so by incorporating gender perspectives when considering the impacts of how the Global Environment Facility decides to fund projects, deeming public transportation projects as a particular priority because they reduce emissions from vehicles and also provide mobility to women living in places where they are not taught to drive (UNFPA, "Facing a Changing World," 27). The GEF has "provided or leveraged more than $40 billion in funding for environmental projects in the developing world since 1991." See UNFPA, "Facing a Changing World," 56.

Into the Darkness with Hearts Ablaze

For Christians, a vision of justice as participation enlivens a theology of mobilization. A theology of mobilization asks every community and all cultures, traditions, and peoples to consider any possible way they may be able to lend their voice and vocation to the collective effort to advance planetary flourishing.[43] When taken in the larger span of time and space, the gross annihilation of life on Earth is unshakably disturbing. It is haunting to think that, in this overwhelmingly vast, dark, and cold cosmos, only one pale blue dot—at least as far as human civilization can say with any certainty at present—is home to the vast array of life-forms that have ever been known and ever may be known by human beings; yet our species persists in systematically plucking this miracle of life out of existence.[44] Sadly, we do this to other species even as we segregate and sometimes annihilate members of our own for the most minuscule of perceived differences.[45] We

---

43. This phrase, "A Theology of Mobilization," is partly in response to, and inspired by, Lester Brown's call to action, and I am indebted to William French who encouraged me to land on a short phrase that encapsulates the primary thrust of my theological project (this phrase was among his suggestions). See Brown, *Plan B 3.0*.

44. Sagan's, now famous, reflection on human history within Earth's cosmic context is not only chilling but resonates with my own argument here, especially when he says, "That's here. That's home. That's us. On it everyone you love, everyone you know, everyone you ever heard of, every human being who ever was, lived out their lives. The aggregate of our joy and suffering, thousands of confident religions, ideologies, and economic doctrines, every hunter and forager, every hero and coward, every creator and destroyer of civilization, every king and peasant, every young couple in love, every mother and father, hopeful child, inventor and explorer, every teacher of morals, every corrupt politician, every 'superstar,' every 'supreme leader,' every saint and sinner in the history of our species lived there—on a mote of dust suspended in a sunbeam." See Sagan, *Pale Blue Dot*, xv–xvi.

45. Certainly, differences between and among individuals, traditions, and cultures exist, as do differences in outlooks and worldviews within such communities over time and across contexts, and I do not think it is helpful to minimize those differences when they contribute to the richness of the human experience. That said, His Holiness the 14th Dalai Lama and many others have long been a champion of what may be considered "our common humanity," in the sense that those "factors which divide us are actually much more superficial than those we share. Despite all the characteristics that differentiate us—race, language, religion, gender, wealth, and many others—we are all equal in terms of our basic humanity." See Dalai Lama XIV Bstan-'dzin-rgya-mtsho, *Beyond Religion*, 29. To this, I humbly add that given the scope of the ecological and climate challenges before us, I wonder if it would not be helpful for humanity to not only consider an emotive appeal to a sense of common humanity, but whether perhaps there may be any emotive appeal to humanity's shared "earthiness" or a sense of "shared being" with all other life forms. For people of Christian or Jewish faith, perhaps a return to the Genesis account of God's one word for all "breathing creatures," as discussed in chapter 3.

are living amidst an extinction event of our own making and we cannot allow ourselves to be paralyzed by the worry, anger, fear, and despair that come with acknowledging the magnitude this problem.

The strange absurdity of all this is that it is within our collective power to alter the trajectory on which we find ourselves fumbling along. We have messed up, yes, but we need not quibble about it endlessly—the time for that is over—now is the time to mobilize into action and do something about it. During such a turbulent time when so much of the human-Earth future hangs in the balance, a theology of mobilization cultivates a creative, all-aboard, all-hands-on-deck, and all-of-the-above approach to problem solving.[46] Every little thing that any one of us can do to name, claim, or open space in the present moment for justice to break in and buy us time, has a role to play in humanity's collective march toward a more just and verdant world. Each contribution is an opportunity to inspire, renew, and sustain the world with a vision of what may be possible. This is the hope on which the promise of our future rests.

Over the long haul, transformation will be necessary across legal, sociopolitical, economic, and market systems. But, this will not be enough and may not happen at all without some recognition of, and attention to, the way human beings are inspired toward action by the kind of stories embodied in many of the world's religions. A theology of mobilization—one with enough thrust to rally Christian faith communities and other people of goodwill into action—recognizes this as part of the way forward and makes use of it for the common good. A theology of mobilization aims to inform, inspire, and mobilize Christian communities to partner with all others in actively imagining and creating a new world wherein the poorest and most vulnerable among us are included in the decisions affecting their lives. Moreover, it names and claims a vision of the world in which the most vulnerable ecosystems and nonhuman members in the community of life are valued too, and not excluded from those decisions made on their behalf affecting their ability to survive and thrive. The only way to journey into the future is together.

A more just and verdant world, as I imagine it, is one in which every body, mind, heart, and soul has an opportunity to thrive while contributing to planetary healing and the flourishing of others. For human bodies

---

46. This references those concerns and solutions of a wide-ranging scope and magnitude, such as represented in Gore's book, including his call for a "Global Marshall Plan." See Gore, *Earth in the Balance*, 297–301.

to flourish, they require access to clean air, water, and food. They require medical care appropriate to each person's gender, age, and stage in life. They require safe and healthy living conditions, including access to the kind of economic opportunities that sustain and support financial health. For minds to flourish, they require access to educational opportunities appropriate to an individual's curiosity, ambition, and intellectual and creative abilities.

For hearts to flourish, they require strong social support networks, such as those created by immediate and extended families, close friendships, vibrant workplaces, religious communities, or ties with those who share interests in hobbies, politics, or a motivation for social activism. They may require also access to professional mental and emotional support services, specifically, for those traumatized by abuse, neglect, war, or by any such experiences that damage bodies, minds, and hearts.

For souls to flourish, the human spirit requires opportunities for leisure, rest, and refreshment. Activities such as prayer, meditation, mindfulness, walking, and quiet contemplation offer reflective opportunities to consider one's relationship with others and ponder one's place in the world. For some, this may be in the mode of reflection on one's relationship with God, with one's deepest self, or with those for whom one lives and loves. It may mean making space and time for the examined life, as opposed to the distracted life many of us lead—a state in which our technologies can relentlessly work against us by isolating us or shortening our attention span and relegating our interest to only the most immediate and pressing of concerns. A more just and verdant world creates opportunities for people to give shape to the way they might imagine their lives as meaningful during the very brief time each of us has on this planet.

For people of Christian faith who share this vision, one question may be how to mobilize to make this imagined world a reality.[47] If human beings

---

47. Although expressly grounded in Christian traditions, and with the intent to help mobilize Christian communities into action, I offer what follows with that hope it inspires others to go and do likewise while working out of their traditions. I intend it in no way to be an exhaustive or comprehensive pathway forward, but rather a starting point or node in ongoing discussions within Christian communities about how to make sense of faith and its traditions in light of contemporary challenges. These activities represent what I see to be broad themes in the traditions that lend themselves to being reclaimed, drawn forward, and applied to some of the many problems confronting the world. But, they also include suggestions for ways Christian ideas can be constructively challenged, so the traditions continue growing, expanding, and evolving as necessary in response to entirely new problems posed, and new insights offered by, contemporary scientific

are "storytelling culture-dwellers," as Oelschlaeger puts it, then as Fasching and DeChant note, "the kind of story we think we are in and the role we see ourselves playing in that story" really does matter.[48] We need good stories and we need stories that link faith with action. Our deeply held values and beliefs about who we are as human beings or as people of faith are meaningless if they are divorced from the way we choose to live in the world. The most robust forms of faith are not the kind that allow us to rest comfortably and confidently in the safety of our predetermined worldviews—rather, they make us uncomfortable by challenging us and opening us up to the inner and outer transformations that will be necessary for us to sustain the labor of building a more just and verdant world. I see three practical actions of faith Christians can take—or three disciplines Christians can embody—to reaffirm, renew, and revive the transformative power of their traditions so they better speak to the needs of the world today.

1. Cultivate a gritty kind of faith

Christian Scriptures begin with a story about the soil and how we are made from and for it. There, beginning with the very first passages of the Bible in Genesis 2 is an invitation for Christians to critically assess their sense of what it means to be human in relation to God, to others, and the world. Genesis 2 is a beautifully told, allegorical story of creation in which the human is oriented in terms of a reverent form of ecological humility. We need more stories like that, stories that place ecological humility at the heart of what it means to be a person made and loved by God. We need stories that temper the long-standing anthropocentric focus dominating many expressions of Christian thought, including the pride of anthropocentrism. We need also stories that do more than critique divisive constructs such as androcentrism, racism, homophobia; we need stories that make valuing some people above others impossible for people of faith.

Christians ought to draw on their ancient biblical stories as a source of inspiration and illumination for a life lived well in the world today. To do so is to read them in ways that are responsive to the signs of the times

---

inquiry. I offer them with my sincerest hope they help mobilize Christians to employ the full force and power of their spiritual and religious traditions. To create and sustain the tremendous energy required by people over the long-haul, folks from all walks of life will need to connect and collaborate with each other in their shared concern for global poverty and ecological degradation.

48. Oelschlaeger, *Caring for Creation*, 9–10; Fasching and DeChant, *Comparative Religious Ethics*, 6.

in which we live. Christians can and must reclaim an ancient sense of what it means to be a person before their God, before others, and before the whole family of creation, but in ways that are informed by what the hard and social sciences tell us about the world and also about what it means to be human. The Scriptures do not fit well in a nice and tidy conceptual box—they are powerful, potentially dangerous stories—they are meant to live and breathe and form and shape a believer's life. Often, they are framed in ways that exclude, separate, and marginalize one group to the benefit of another. At best, they can transform us by opening us up to the great needs of the world.

Living out a kind of faith that opens us up, rather than shuts us down, requires a good deal of grit. In a world plagued by disproportionate overconsumption, many of us are going to have to learn how to scale way back on the way we have been living. We will have to acknowledge that a "throwaway" culture is a major part of what is wrong with the world today, and that the way too many of us live our lives feeds the monster that is quite literally devouring this magnificent, precious world. How does one kneel humbly at the feet of God in prayer and worship when one refuses to even consider the way one lives in the world, and whether it is simple, generous, and just?

Our bodies matter; we have needs that must be met. But those needs are satiable—only the desire for more is insatiable. Self-worth does not come from material acquisition. Christians believe people are made and loved by God, for lives to be lived with and for others. This is what it means to be part of a community of mutual responsibility—to look inward and ask ourselves how we might turn outward and live differently. Nobody should expect it is easy to change the way one has always been doing things. Faith lived out like this takes the courage to hope amidst discouraging forms of despair, tenacity in the face of seemingly insurmountable challenges, and a resiliency of spirit to keep on trying to live for others even when we find ourselves stubbornly clinging to our own narrow self-interest. Lived out like this, faith takes grit.

2. Develop a more informed sense of creation's inherent worth, an appreciation for its intrinsic goodness, and imagine new ways to celebrate it

Scientific insights have revealed so much more about the mysteries of the Universe we inhabit and the Earth we call home than at any other time in the history of civilizations. This includes a growing understanding of global phenomena like weather and climate, plate tectonics and continental drift, as well as forces like natural selection, Mendelian inheritance, and the evolution of species. We continue to grow in our knowledge of ecological processes, and the complex interactions between biogeochemical cycles, population dynamics, and the general activity of life on and in Earth's life support systems. Massive, planetary-wide problems such as ocean acidification and coral bleaching, desertification, and species extinction are part of this story too.

As our collective human power and presence on the Earth grows, the need for careful, judicious, and humble wielding of this newfound power also increases. It could be a frightening display of hubris to act as though we can control the trajectory of something like climate change through activities such as geoengineering, even in the name of "saving the planet," when the primary consequence of human activity has so far resulted in what some are calling the sixth major mass extinction the planet has ever seen. As we strive to better understand how Earth's systems work, we can employ that knowledge to promote human flourishing and to live more sustainably. But, scientific knowledge and human understanding of Earth and its processes has so far not yet yielded, for lack of a better word, love for the Earth.

People of faith must retrieve a theology of creation in which respect and appreciation for the Earth as its own agent before God is celebrated. This means learning to see the world through a lens of awe and wonder. It means learning to live in a way that bears witness to ecologically complex relationships of mutual dependency. Our best science is not an adversary in this pursuit. It is an ally. The human person is a biologically embedded, interconnected being in an ecological reality with needs and limitations in relation to the needs and limits of the larger planetary biome. As living, breathing, embodied creatures with hearts and minds, we have physical, mental, and emotional needs that must be met. But the Earth, our source of life and sustenance has ecological limits. We need to learn to live more equitably within them. We also need stories that arouse our love and respect for the Earth's own inherent worth, as well as stories of Creation's dignity and standing before God. Most importantly, we need to live out and lean in to those stories that depict this reality for us, and help us bring it into being.

> 3. *Create spaces to promote equity, diversity, participation, and an enriched theological vocabulary*

A church with a richer diversity of voices has a better theological vocabulary. The word "theology" is derived from two Greek words that together mean, quite simply, "words about God."⁴⁹ What it means to do Christian theology can certainly be more nuanced, and it has quite often become much more complicated. Still *lo cotidiano*, or the experiences of everyday life, are foundational for what it means to do Christian theology and ethics. So many people are precluded from greater participation in society and the life of the church because of various forms of poverty and social and economic inequalities.

Christians following in the ministry modeled by Jesus need to expand this space as much as possible so it includes and sustains previously marginalized voices like these. But, the spaces of *whose* experiences are counted needs to be expanded in other ways too. The prevailing perspectives customarily recognized by the church do not nearly represent the incredibly rich diversity of human beings and their experiences—our various ways of seeing, loving, and being—that make up the differentiated threads in the larger tapestry of human living.⁵⁰ Soliciting diverse perspectives, insisting on equity, and inviting participation in all aspects of community life create the spaces necessary for faith to come alive and be animated by action.

By the same token, could the expansion of this space somehow include making more room for what the cosmos reveals about the nature of God? Moving forward, can Christians make a stronger effort to recognize a special role for ecological and evolutionary Earth sciences in the formation of new spiritual-religious creation stories? Can Christians learn to imagine emerging scientific stories of the Universe as a kind of dialogue partner in creating entirely new words about God? I think so; at least, I am hopeful this can continue to happen within the various iterations of Christian faith that have already begun embracing the sciences. Christians so desperately need new words about God that not only help them make better sense of

---

49. McGrath, *Christian Theology*, 86.

50. In other words, one question is how Christian faith communities might include a more diverse set of life experiences that lead to different perspectives about God. Seeking out those various perspectives in a way that is more responsive to a contemporary social and ecological context, and namely, one that values both human dignity through social justice and also more-than-human dignity through a robust practice of ecological responsibility, may be one way to critically assess them in relation to the traditions. On the "distinctive but interconnected" relationship between social location and ecological location, see Daniel Spencer's description. Also noteworthy is how Spencer builds on feminist and liberation theological reflections on ecological ethics by including a lesbian and gay theo-ethical reflection to build what he calls an "erotic ethic of eco-justice." See Spencer, *Gay and Gaia*, 295–96, 339–45.

what it means to be human and to live more equitably with each other but also how to live in the world as we now know it.

Words spoken about God out of love for the world in Christian liturgy, in prayer, and in the general life of many Christian faith communities are germinating, but they have yet to fully blossom. The church needs more words about God inspired by earthy forms of sacramentalism and enfleshed spiritualities. It also needs new forms of prayer and ritual grounded in the cycles of Earth and the story of the Universe, that invoke and embolden new ways of being in the world. Christians who are already doing this work often start from the rich sources of stories embedded in their traditions, but there is ample room to expand this garden of Christian faith in a spirit of openness and creativity.[51] We need more experimentation, trial and error, and freedom in the cultivation of world-aware spiritual practices as they bubble up from the lived experiences of people in the pews. A vibrant church has a door that is always open to the living faith of its people.

People like Sister Mariam MacGillis, a Catholic Dominican Sister and founder of Genesis Farm in New Jersey, are doing this. She and others have found ways to cultivate practices that contextualize Christian spirituality in terms of a sacred earth community and cosmic liturgy. At Genesis Farm, guests can learn forms of body prayer, tracing the four cardinal directions at sunrise, or walking through Stations of the Earth that evoke "the notion of the earth's Passion, providing space for confession or reconciliation, as well as atonement for human sins perpetuated against the planet."[52] As Christians continue to learn new words about the world and the cosmos from the hard and social sciences, imagine how that might expand and enrich the way they talk about God. Imagine all the new stories they will tell, the inner and outer transformations they will provoke, and solidarity with Earth and it's healing that they will yield.

Be gritty. Love the world. Make justice happen. These three disciplines emerge from and function in response to the primary undergirding principle I have endeavored to clarify in this book: Concern for the world, for people and for planet—or concern for social justice and concern for the Earth—are two sides of the same coin. If one without the other was ever

---

51. Few other sources are as responsible for my sense of optimism about what is possible than those found in the work of these people: Tucker and Grim, *Worldviews and Ecology*; Tucker, *Worldly Wonder*; Grim and Tucker, *Ecology and Religion*. Also, Swearer and McGarry, eds., *Ecologies of Human Flourishing*; Lodge and Hamlin, eds., *Religion and New Ecology*.

52. Taylor, *Green Sisters*, 243.

viable, that is certainly now no longer the case. I have aimed for realistic, down-to-earth, and practical ways for Christians to carry their robust traditions into a new future that contributes to shared human and planetary flourishing. Openness to new responses and new insights is requisite for authentic dialogue between traditions and across different ways of knowing and seeing the world. Christians can and should be anticipating how new moral problems and new scientific insights might become fodder for new manifestations of a robust Christian moral imagination. We can integrate social justice with ecological concern. We can integrate retrieval with reconstruction. We can integrate ethics and moral contemplation with a foundation of facts rooted in the hard and social sciences. We need and we have a galvanizing call to action. All are vital parts of a theology of mobilization for the twenty-first century.

SEEDS OF CHANGE

Throughout this book, I have referenced another core concern in Christian ethics. Its aim is to ground theological reflection by asking, simply: So what? Or rather, how does any of this matter for those whose lives are lived at the margins? The "so what?" question exists to evaluate the ethical thrust of any theological excursus, because academic conversations tend too often to get bogged down in so much theory that they lose any ability to actually impact the world at all.[53] The urgency of both global poverty and planetary degradation warrants the need for intense pressure on theologians, clergy, and lay leaders who simply ought not to waste time in unnecessarily complicating what is essentially a clear and compelling call to action.[54] Urgency and a shared call to action are partly what drive the need for a theology of mobilization. At the end of the day, the precise way in which humanity shifts the present paradigm of human-Earth relations may matter less than

---

53. I am thinking here of James Garvey's argument that sometimes it is "the whole of a life and the way it's lived" that is the more appropriate question in ethics than is the particular moral dilemmas surrounding individual acts, especially within the context of a challenge like climate change in which impactful individual action to thwart its consequences may be impossible. See Garvey, *Ethics of Climate Change*, 147–51.

54. For an example of the great diversity of ethical arguments that ultimately rally around a push for action on issues like climate change, see Northcott, *Political Theology*; and Northcott, *Moral Climate*. Also, Gardiner, *Perfect Moral Storm*; Nanda, ed. *Climate Change*; Thompson and Bendik-Keymer, eds., *Ethical Adaptation*; Coward and Hurka, *Greenhouse Effect*; and Harris and Vanderheiden, *Ethics and Environmental Policy*.

that we all rally together and push. I am hopeful that if we make a concerted effort, we will figure out how to conscientiously and collectively engage the challenges before us—by rallying together in co-mission to preserve, sustain, and enhance life on this "pale blue dot" we call home.

The seeds of change within the legal, sociopolitical, and moral imagination have been sprouting for at least a generation. Changes within market systems, and changes in the way economic value is determined, have been much slower in coming. But, those seeds have certainly been planted. The desire for change has been fomenting in different forms around the world. The beginnings of a theology of mobilization have already taken hold both within Christian communities across the planet and across other religious traditions as well. Many of the world's people are mobilizing and actively working to build this new world in ways that embody some of the vision and activities described in this chapter. What this looks like on the ground is both immediate and practical but also long-term and big-picture in its various iterations. The great work before us is so great in part because it is integrative, collective, and structural.[55] It is also an immense challenge.

Integrative, collective, structural change can prune back the roots of injustice and environmental degradation that have taken hold within and across society. But, we need a model that is wide-ranging, sustainable, and creates a "safe and just space for humanity."[56] It must insist on a social foundation prioritizing human health, food, water, income, education,

---

55. I have elsewhere noted the importance of individual action, specifically within the context of "subsidiarity," and how individual action can be cumulative especially when it is magnified across an individual's various spheres of influence (for example, individuals can take different actions with various effects, within several spheres of influence: household, congregation, city, state, nation, region). See Mastaler, "Case Study on Climate," 65–87, 84–86. For more on subsidiarity as a concept in models of governance, see Colombo, ed. *Subsidiarity Governance*. On the concept of subsidiarity in Roman Catholic social theory, see Grasso, Bradley, and Hunt, eds., *Catholicism, Liberalism, and Communitarianism*, 23–26, 81–96. Also, John Paul II, *Encyclical Letter*, paragraphs 13, 21; and Pope John XXII, *Mater Et Magistra*, paragraphs 53–54, 152; Pope John XXIII, "*Pacem in Terris*," paragraphs 140–41; both in O'Brien and Shannon, eds., *Catholic Social Thought*.

56. See Rockström et al., "Safe Operating Space," 472–5; Rockström et al., "Planetary Boundaries," 1–33. Also, see the excellent work coming out of the Stockholm Environment Institute. Kate Raworth translated this concept for the advocacy community during her tenure with Oxfam International, who in turn helped introduce it to the United Nations with the image of a doughnut. See Raworth, *Safe and Just Space*, 15; and United Nations General Assembly, "Report of the United Nations."

resilience, voice, jobs, energy, social equity, and gender equality.[57] There are basic thresholds in each of these categories every person deserves. Our model also needs to recognize boundaries defined by climate change, freshwater use, nitrogen and phosphorus cycles, ocean acidification, chemical pollution, atmospheric aerosol loading, ozone depletion, biodiversity loss, and land use change.[58] The "sweet spot" for inclusive and just sustainable development is the space between the thresholds necessary for human flourishing and the boundaries of environmental accommodation.[59] Such a model exists, but we need to implement it. The transformation of the world along these lines is how we begin to bend the long arc of the moral universe toward justice for the poorest of the poor and for the Earth.[60]

The great work of transforming human-Earth relations is a work that may be imagined as a process already-in-motion-yet-not-now-fully-realized. The Earth Charter is a significant framework pointing to this yet-to-be and only partly now characteristic of the movement. It is a comprehensive and distinctive document that has been called "the most negotiated document in human history."[61] The vision and principles set forth in the Charter, launched on June 29, 2000 by the Earth Charter Commission, embody a "global ethic" that unites the core themes of "respect and care for the community of life, ecological integrity, social and economic justice, democracy, nonviolence, and peace."[62] These principles offer important ethical scaf-

57. Rockström et al., "Safe Operating Space," 472–5.
58. Rockström et al., "Safe Operating Space," 472–5.
59. With regard to climate change, climate-induced displacement, and international negotiations, there are several concerns I have chosen to omit at this juncture in the name of clarity, and to emphasize a more holistic and less political tone. Conceptually organizing the environmental impact of climate change and the social impact of climate-induced displacement within this larger framework of moral concern may be the only way in which a fair and just climate response may ultimately be negotiated and achieved. For a general overview of how climate accords are negotiated within the context of international relations, see Luterbacher and Sprinz, eds., *International Relations*.
60. Though various iterations of this aphorism, the moral arc of the universe bending toward justice, have been used throughout the late nineteenth and twentieth centuries, I attribute my use of the phrase here to the Rev. Dr. Martin Luther King Jr.
61. John Lane quotes Steven C. Rockefeller, whom Lane observes is one of the key architects of the document, with this description. See Lane, "Lake Conestee," 66. For a helpful overview of the Earth Charter and its significance, see Bosselmann and Engel, *Earth Charter*; and Westra and Vilela, eds., *Earth Charter*.
62. See Earth Charter Initiative, "*Earth Charter.*" For a general introduction to the concept of a "global ethic" in relation to the Earth Charter, see Westra, *Living in Integrity*; and Miller and Westra, *Just Ecological Integrity*. Also Dower, " Global Ethics," 15–28;

folding upon which current and future generations may live out the most significant paradigm shift in the collective cultural history of our species.[63] Many of the Earth Charter's ideas resonate with some of the most deeply held ethical values and beliefs exemplified across faith traditions, including Christian traditions. The document's potential to inspire and mobilize the world into action has almost certainly not yet reached its fullest potential to galvanize an emerging planetary civilization. For now it remains merely an aspirational document, but it is, perhaps, one of the most promising seeds of change—the roots of which have only just begun to penetrate the firmaments of humanity's most imposing social systems. The day it unfurls its first true leaves may be one future generations recall as the day humanity 'woke.'

HOPE FOR THE JOURNEY

I cannot help but choose to believe—or rather, I suppose I choose to hope—that human beings are their most creative, their most collaborative, and their highest and best selves when they tap into whatever great and mysterious source inspires and sustains them. I am talking about that enduring human ability to persevere in the direst of circumstances and to persist amidst all forms of despair. If Christians are able to turn to their traditions, and create new stories that help them live out a renewed sense of ecological humility, a sense of creation's inherent worth and intrinsic goodness, and respect for lived experiences as an ongoing important source for theological ethics, then I think we can legitimately anticipate the kind of change the world needs. If Christians are able to make peace with the evolutionary and ecological sciences, and see them for the allies they really are, then they may actually be able to begin harnessing the enduring influence of their traditions in ways that constructively accompany the poorest of the poor, human or otherwise, in the pursuit of shared flourishing.

The call to action found in the closing words of the Earth Charter proposes that "ours be a time remembered for the awakening of a new reverence for life, the firm resolve to achieve sustainability, the quickening

---

Engel, "Covenant Model," 29–46; Rockefeller, "Global Ethics," 1–7; and Rockefeller, "Crafting Principles," 3–23.

63. For examples of how the world's religions inspire and facilitate the Earth Charter, see Tucker, "World Religions," 115–28; and Tucker, *Worldly Wonder*; Also, Ruether, *Integrating Ecofeminism*.

of the struggle for justice and peace, and the joyful celebration of life."[64] Sit with that for a moment. Let it move you. Then consider what you can do about it. People of faith, will you rise up and join with others in bold courage to quicken the struggle for justice and peace? Will you uncover the will and discover the resolve to chart a way forward in this dark and trying time? Will you resist the shadows of despair and earnestly resolve to set forth in co-mission to preserve, sustain, and enhance life on this planet we call home? Will you commit to working together with others, mobilizing to create this new world, ever so pregnant with the possibility of shared planetary flourishing? In short, do you have the fire in your veins to answer this call and step into the darkness with unwithered hope and heart ablaze?

This is far from where we began. I began by asking a very different set of questions: Why is it that those who are trained in ethical reasoning, either professionally or via a religious tradition or both, have typically not spent much time on ecological concerns relative to other moral dilemmas emerging from human experience, like birth and death, sexuality, or war? More specifically, why is it that profound environmental and social challenges such as environmental degradation and climate change, have yet to receive adequate attention from the world's religious traditions in a way that helps to mobilize human civilization around a moral call to action on the part of planetary survival? What is it about the way certain faith communities, and Christian communities in particular, paint a picture of the world and understand the role of the human person in a way that either hinders them or compels them to work for collective, structural change? What kind of change is necessary to sustain the vitality of Earth's ecosystems and some of the most vulnerable populations directly dependent upon them? What, if any, progress might offer a basis for hope? In grappling with these questions earlier in the book, and now contemplating a call to action, I hope you walk away inspired by the most important characteristic of a theology of mobilization: the human community can act, we ought to act, and the time to act is now.

But does this mean we *will* act? No, unfortunately, it does not. That is the base concern. Can we act, and will we act are two different questions. I hope we act, but we will have to overcome the paralyzing effect that attends the realization of what is happening. Hope is probably the most powerful tool we have in the shed right now. How do we maintain an attitude of hopefulness, which is empowering, while still acknowledging the madness

---

64. Earth Charter Initiative, *"Earth Charter."*

that is our collective assault on this planet? I am sometimes asked whether anyone can be truly hopeful, often by those who quite understandably find themselves feeling exasperated, overwhelmed, or totally disheartened by the immensity and severity of the problem before us. Often, the question is quite direct and personal, as I am asked, "James, are *you* hopeful? I mean, are you personally *really* hopeful? And if you are, what could possibly give you hope?" While I have pointed to several reasons why I think it is quite reasonable for us to be hopeful, namely because our species is capable of the kind of mobilizing change human civilization needs to make, I have also argued that stories often weigh more heavily on hearts and minds than argument alone. In that vein, I offer a personal story to explain my own reason for hope amidst so much despair.

It was late spring, shortly after returning from my tenure in rural Bangladesh, and before beginning my doctoral studies. I was with my extended family at a secluded cabin in the Appalachian Mountains. My first nephew was learning how to walk and talk and I was thrilled to be able to spend some time with him in such a beautiful natural environment. One afternoon while picnicking in a small park at the base of the mountains, my two-year-old nephew and I stumbled across a field of dandelions whose flowers were spent, leaving an almost endless sea of white, puffy seed-heads ready to disperse at the slightest disruption. Having grown up in and around open spaces my whole life I always find these sights quite spectacular.

Instead of encouraging my nephew to run through the field and set-off a blizzard of seeds, which certainly would have been delightful in its own way, we crouched down near the edge of it and picked one dandelion. I remember the way his delicate little fingers carefully held that one small seed-head before I encouraged him to blow on it gently in order to see what happens. When he did, I watched his eyes light up with unbridled astonishment as the seeds flew into the air and floated upward like magic. His quiet response was but a whisper and barely audible, but it has stuck with me. It was the only word he knew to express the sense of awe and wonder he was experiencing. He stared with wide-eyes and pointed before letting out a very drawn-out "wow." That word trailed off almost as quietly as it had begun. In that "wow" I find tremendous hope. So much hope.

When I begin to feel overwhelmed by the great harm we are doing to each other and to the Earth's ecological systems, I try to make room in the busyness of my daily schedule to seek out those moments in which I might recapture something of that sense of awe and wonder. It is a gift to be able to stop and marvel, even if only for a moment, at the beauty and grandeur that makes up this cosmic journey of life on our planet. I find these moments lend humility and clarity by grounding my work in a larger trajectory of time—a trajectory that transcends my own life and situates me in a community of people who have long worked, and will long continue the work in this effort to nurture the miracle of life that has emerged on our planet. That inspires me to keep on going, even when it sometimes looks like our efforts are flying in the face of insurmountable odds. I so desperately wish more of us would see and engage the world in this way—to more clearly and more often experience that deeply mysterious and seductive sense of joy and peace that really is all around us. We only have to look for it. Seek God in all things, through the microscope, the telescope, and everywhere in between.

The same energy that brought the cosmos into being, sustains it still; it fills your lungs, it courses through your veins, and it animates your spirit. There is hope in that.

Bibliography

Abend, Lisa. "In Spain, Human Rights for Apes." *Time*, July 18, 2008. http://content.time.com/time/world/article/0,8599,1824206,00.html (accessed September 4, 2018).

Abeygunawardena, Piya, et al. *Poverty and Climate Change: Reducing the Vulnerability of the Poor through Adaptation*. Washington, DC: The World Bank, 2009.

Agesa, Jacqueline, and Richard U. Agesa. "Gender Differences in the Incidence of Rural to Urban Migration: Evidence from Kenya." *Journal of Development Studies* 35, no. 6 (1999) 36–58.

Albanese, Catherine L. *Nature Religion in America: From the Algonkian Indians to the New Age*. Chicago: University of Chicago Press, 1990.

Alkire, Sabina, and Maria Emma Santos. "Multidimensional Poverty Index." In *OPHI: Oxford Poverty & Human Development Initiative*, 1–8. Oxford: University of Oxford, July 2010.

American Baptist Churches USA. "Creation and the Covenant of Caring." In *Our Only Home: Planet Earth*, edited by Owen D. Owens, 34–39. Valley Forge, PA: American Baptist Churches, 1991.

Ammer, Margit. "Climate Change and Human Rights: The Status of Climate Refugees in Europe." In *Protecting Dignity: An Agenda for Human Rights*, edited by Mary Robinson and Paulo Pinheiro, 1–74. Geneva: Swiss Federal Department of Foreign Affairs, 2009.

Amstutz, Mark R. *International Ethics: Concepts, Theories, and Cases in Global Politics*. 4th ed. Lanham: Rowman & Littlefield, 2013.

Anderegg, William R. L. et al. "Expert Credibility in Climate Change." *Proceedings of the National Academy of Sciences* 107, no. 27 (June 2010) 12107–12109.

Anderson, Bernhard W. *From Creation to New Creation: Old Testament Perspectives*. Minneapolis: Fortress, 1994.

Andrews, Molly. *Shaping History: Narratives of Political Change*. Cambridge; New York: Cambridge University Press, 2007.

An-na'im, Abdullahi A. "The Politics of Religion and the Morality of Globalization." In *Religion in Global Civil Society*, edited by Mark Juergensmeyer, 23–48. Oxford: Oxford University Press, 2005.

Appleby, Joyce Oldham. *The Relentless Revolution: A History of Capitalism*. New York, NY: W. W. Norton, 2010.

Aquinas, Thomas. *Summa Theologiae*. Translated by Paul E. Sigmund. New York: W. W. Norton, 1987.

Bibliography

———. *On Truth (De Veritate)*. Translated by Mulligan, SJ, from Leonine Edition. Chicago: Henry Regnery, 1952.

———. *On Kingship: To the King of Cypress*. Translated by Gerald B. Phelan. Toronto: Pontifical Institute of Medieval Studies, 1949.

———. *Summa Contra Gentiles*. Translated by English Dominican Fathers from Leonine Edition. New York: Benziger Brothers, 1924a.

———. *Summa Theologiae*. Translated by Fathers of the English Dominican Province. Vol. 5. Westminster: Christian Classics, 1948.

———. *Summa Theologiae*. Translated by English Dominican Fathers from Leonine Edition. New York: Benziger Brothers, 1924b.

———. *The Summa Theologica*. Translated by Fathers of the English Dominican Province. Westminster, MD: Christian Classics, 1981.

Aquino, María Pilar. "Theological Method in U.S. Latino/a Theology: Toward an Intercultural Theology for the Third Millennium." In *From the Heart of our People: Latino/a Explorations in Catholic Systematic Theology*, edited by Orlando O. Espín and Miguel H. Díaz, 6–48. Maryknoll: Orbis, 1999.

———. *Our Cry for Life: Feminist Theology from Latin America*. Maryknoll: Orbis, 1993.

Aristotle. *Nicomachean Ethics*. Translated by Martin Ostwald. Indianapolis: Bobbs-Merrill, 1962.

Armitage, Derek. "Adaptive Capacity and Community-Based Natural Resource Management." *Environmental Management*. 35, no. 6 (2005) 703–15.

Associated Press. "Germany Guarantees Animal Rights in Constitution." *USA Today*, May 18, 2002. https://usatoday30.usatoday.com/news/world/2002/05/18/germany-rights.htm (accessed September 4, 2018).

Atran, Scott. *In Gods we Trust: The Evolutionary Landscape of Religion*. Oxford: Oxford University Press, 2002.

Augustine. *Concerning the City of God Against the Pagans*. Translated by Henry Bettenson. London: Penguin, 2003.

———. *The Confessions of St. Augustine*. Translated by John K. Ryan. Garden City, NY: Image, 1960.

———. *The Trinity*. Translated by Stephen McKenna. Washington, DC: Catholic University of America Press, 1963.

Bain, Paul G., et al. "Promoting Pro-Environmental Action in Climate Change Deniers." *Nature Climate Change* 2, no. 8 (August 2012) 600–603.

Baird Callicott, J. "The Worldview Concept and Aldo Leopold's Project of 'World View' Remediation." *Journal for the Study of Religion, Nature & Culture* 5, no. 4 (2011) 510–28.

Baldwin, John W. *The Scholastic Culture of the Middle Ages 1000–1300*. Long Grove, IL: Waveland, 1997.

Barkley Rosser, J. "Belief: Its Role in Economic Thought and Action." *American Journal of Economics and Sociology* 52, no. 3 (1993) 355–58.

Barnosky, Anthony D., et al. "Has the Earth's Sixth Mass Extinction Already Arrived?" *Nature* 471, no. 7336 (2011) 51–57.

Bauman, Whitney. "Consumerism and Capitalism: The True Costs of Integrity." *Dialog* 49, no. 4 (Winter 2010) 263–64.

Bavington, Dean. "*Homo administrator*." In *Recognizing the Autonomy of Nature: Theory and Practice*, edited by Thomas Heyd, 121–36. New York: Columbia University Press, 2005.

Bibliography

Becker, Marc. "Correa, Indigenous Movements, and the Writing of a New Constitution in Ecuador." *Latin American Perspectives* 38, no. 1 (2011) 47–62.
BeDuhn, Jason. *Augustine's Manichaean Dilemma: Conversion and Apostasy, 373–88 C.E.* Vol. 1. Philadelphia: University of Pennsylvania Press, 2010.
———. *The Manichaean Body: In Discipline and Ritual.* Baltimore: Johns Hopkins University Press, 2000.
Bell, Daniel. *The Cultural Contradictions of Capitalism.* New York: Basic, 1976.
Bellah, Robert N. *Habits of the Heart: Individualism and Commitment in American Life.* Berkeley: University of California Press, 1985.
Benzoni, Francisco J. *Ecological Ethics and the Human Soul: Aquinas, Whitehead, and the Metaphysics of Value.* Notre Dame: University of Notre Dame Press, 2007.
Berger, Peter. "Religion and Global Civil Society." In *Religion in Global Civil Society*, edited by Mark Juergensmeyer, 11–22. Oxford: Oxford University Press, 2005.
———. *The Sacred Canopy: Elements of a Sociological Theory of Religion.* Garden City, NY: Doubleday, 1967.
Berger, Peter, and Thomas Luckmann. *The Social Construction of Reality: A Treatise in the Sociology of Knowledge.* Garden City, NY: Doubleday, 1966.
Berry, Thomas. *Buddhism.* New York: Hawthorn, 1967.
———. *The Dream of the Earth.* San Francisco: Sierra Club, 1988.
———. *Evening Thoughts: Reflecting on Earth as Sacred Community.* Edited by Mary Evelyn Tucker. San Francisco: Sierra Club, 2006.
———. *The Great Work: Our Way into the Future.* New York: Random House, 1999.
———. "The Historical Theory of Giambattista Vico." PhD diss., Catholic University of America Press, 1949.
———. *The New Story.* Chambersburg: Anima, 1978.
———. *Religions of India: Hinduism, Yoga, Buddhism.* Chambersburg: Anima, 1992.
———. *The Sacred Universe: Earth, Spirituality, and Religion in the Twenty-First Century.* Edited by Mary Evelyn Tucker. New York: Columbia University Press, 2009.
Berry, Wendell. "The Way of Ignorance." In *The Virtues of Ignorance: Complexity, Sustainability, and the Limits of Knowledge*, edited by Bill Vitek and Wes Jackson, 37–50. Lexington: University Press of Kentucky, 2008.
Bhubaneswar, Orissa, and Jana Sanjaya. "Cholera Death Toll in India Rises." *BBC News*, August 29, 2007. http://news.bbc.co.uk/2/hi/south_asia/6968281.stm (accessed September 4, 2018).
Bieringer, Reimund, and Mary Elsbernd, eds. *Normativity of the Future: Reading Biblical and Other Authoritative Texts in an Eschatological Perspective.* Leuven: Peeters, 2010.
Bishop, Joshua. *The Economics of Ecosystems and Biodiversity in Business and Enterprise.* New York: EarthScan, 2012.
Black, Richard. "Big Profit from Nature Protection." *BBC News*. http://news.bbc.co.uk/2/hi/8357723.stm (accessed September 4, 2018).
Bodansky, Daniel. "The United Nations Framework Convention on Climate Change: A Commentary." *Yale Journal of International Law* 18, no. 2 (1993) 451–558.
Bodansky, Daniel, and Lavanya Rajamani. "The Evolution and Governance Architecture of the Climate Change Regime." In *International Relations and Global Climate Change*, edited by Detlef Sprinz and Urs Luterbacher. 2nd ed. Cambridge, MA: MIT Press, 2016. https://ssrn.com/abstract=2168859

Bibliography

Boff, Leonardo. "Liberation Theology and Ecology." In *Money & Faith: The Search for Enough*, edited by Michael Schut, 134–39. Denver: Morehouse Education Resources, 2008.

Bonner, Gerald. *St. Augustine of Hippo: Life and Controversies*. Philadelphia: Westminster Press, 1963.

Bosselmann, Klaus, and J. Ronald Engel. *The Earth Charter: A Framework for Global Governance*. Amsterdam: KIT, 2010.

Bouma-Prediger, Steven. *For the Beauty of the Earth: A Christian Vision for Creation Care*. Grand Rapids: Baker Academic, 2010.

———. *The Greening of Theology: The Ecological Models of Rosemary Radford Ruether, Joseph Sittler, and Jürgen Moltmann*. Atlanta: Scholars, 1995.

Boyd, Robynne. "One Footprint at a Time." *Scientific American*. http://blogs.scientificamerican.com/plugged-in/2011/07/14/one-footprint-at-a-time/ (accessed September 4, 2018).

Breube, Alan, and Bruce Katz. *Katrina's Window Confronting Concentrated Poverty Across America*. Washington, DC: Brookings Institution, 2005.

Brock, Rita Nakashima, and Rebecca Ann Parker. *Saving Paradise: How Christianity Traded Love of this World for Crucifixion and Empire*. Boston: Beacon, 2008.

Brooks, Bradley. "Pope Speaks Out on Amazon during Brazil Trip." *The Associated Press*. http://www.thestar.com/news/world/2013/07/27/in_brazil_pope_francis_speaks_out_on_the_amazon_environment_and_indigenous_people.html (accessed September 4, 2018).

Brooks, Nick, et al. "The Determinants of Vulnerability and Adaptive Capacity at the National Level and the Implications for Adaptation." *Global Environmental Change* 15, no. 2 (2005) 151–63.

Brouwer, Roy, et al. "Socioeconomic Vulnerability and Adaptation to Environmental Risk: A Case Study of Climate Change and Flooding in Bangladesh." *Risk Analysis* 27, no. 2 (2007) 313–26.

Brown, Lester R. *Plan B 3.0: Mobilizing to Save Civilization*. New York: W. W. Norton & Company, 2008a.

———. *Plan B 4.0: Mobilizing to Save Civilization*. New York: W. W. Norton & Company, 2009.

Brown, Oli. *Migration and Climate Change*. Geneva: International Organization for Migration, 2008b.

Brown, Peter. *Augustine of Hippo: A Biography*. Berkeley: University of California Press, 1967.

Browning, Melissa. *Risky Marriage: HIV and Intimate Relationships in Tanzania*. Lanham: Lexington, 2013.

Bruteau, Beatrice. *God's Ecstasy: The Creation of a Self-Creating World*. New York: Crossroad, 1997.

Bstan-'dzin-rgya-mtsho, Dalai Lama XIV. *Beyond Religion: Ethics for a Whole World*. Boston: Houghton Mifflin Harcourt, 2011.

Butler, Judith. *Gender Trouble: Feminism and the Subversion of Identity*. New York: Routledge, 1990.

———. *Undoing Gender*. New York: Routledge, 2004.

Cahill, Lisa Sowle. *Sex, Gender & Christian Ethics*. Cambridge: Cambridge University Press, 1996.

Bibliography

Cannon, Terry. "Gender and Climate Hazards in Bangladesh." In *Gender, Development, and Climate Change*, edited by Rachel Masika, 45–50. Oxford: Oxfam, 2002.

Carr, David M. "Genesis." In *The Oxford Encyclopedia of the Books of the Bible*, edited by Michael D. Coogan, 316–34. Oxford: Oxford University Press, 2011.

Castro, Alfonso Peter, Dan Taylor, and David Brokensha. *Climate Change and Threatened Communities: Vulnerability, Capacity, and Action*. Rugby, Warwickshire: Practical Action, 2012.

Castro, Victoria, and Carlos Aldunate. "Sacred Mountains in the Highlands of the South-Central Andes." *Mountain Research and Development* 23, no. 1 (February, 2003) 73–9.

Cavalieri, Paola, and Peter Singer. *The Great Ape Project: Equality Beyond Humanity*. New York: St. Martin's, 1994.

Cavanaugh, William T. *Being Consumed: Economics and Christian Desire*. Grand Rapids: William B. Eerdmans, 2008.

Chambers, Robert. *Vestiges of the Natural History of Creation*. 12th ed. London: W. & R. Chambers, 1884.

Chapman, Paul. "Entire Nation of Kiribati to be Relocated Over Rising Sea Level Threat: The Low-Lying Pacific Nation of Kiribati is Negotiating to Buy Land in Fiji so it can Relocate Islanders Under Threat from Rising Sea Levels." *The Telegraph*, March 7, 2012. https://www.telegraph.co.uk/news/worldnews/australiaandthepacific/kiribati/9127576/Entire-nation-of-Kiribati-to-be-relocated-over-rising-sea-level-threat.html (accessed September 4, 2018).

Cheney, Jim. "'The Waters of Separation': Myth and Ritual in Annie Dillard's Pilgrim at Tinker Creek." *Journal of Feminist Studies in Religion* 6, no. 1 (03/01, 1990) 41–63.

Chilton, Paul, and Christina Schaffner. *Politics as Talk and Text: Analytic Approaches to Political Discourse*. Amsterdam: John Benjamins, 2002.

Clark, Gregory. *A Farewell to Alms: A Brief Economic History of the World*. Princeton: Princeton University Press, 2007.

Clements, John M., et al. "Green Christians? An Empirical Examination of Environmental Concern within the U.S. General Public." *Organization & Environment* 27, no. 1 (2014) 85–102.

Cobb, John B. *A Christian Natural Theology: Based on the Thought of Alfred North Whitehead*. Philadelphia: Westminster, 1965.

Cohen, Adam. "What's Next in the Law? The Unalienable Rights of Chimps." *The New York Times*, July 14, 2008. https://www.nytimes.com/2008/07/14/opinion/14mon4.html (accessed September 4, 2018).

Colombo, Alessandro, ed. *Subsidiarity Governance: Theoretical and Empirical Models*. New York: Palgrave Macmillan, 2012.

Cone, James H. *A Black Theology of Liberation*. Maryknoll: Orbis, 1999.

———. *A Black Theology of Liberation*. Philadelphia: Lippincott, 1970.

Connolly, William E. "Capitalism, Christianity, America: Rethinking the Issues." *Political Theology* 12, no. 2 (April, 2011) 226–36.

———. *Capitalism and Christianity, American Style*. Durham: Duke University Press, 2008.

Conradie, Ernst M., and Willis Jenkins, eds. *Ecology and Christian Soteriology* [Special issue of *Worldviews: Global Religions, Culture, and Ecology*] 14, no. 2–3 (2010).

Coote, Robert B., and David Robert Ord. *In the Beginning: Creation and the Priestly History*. Minneapolis: Fortress, 1991.

Bibliography

Copeland, M. Shawn. *Enfleshing Freedom: Body, Race, and Being.* Minneapolis: Fortress, 2009.

Costanza, Robert, et al. "The Value of the World's Ecosystem Services and Natural Capital." *Nature* 387, no. 6630 (1997) 253–60.

Coward, Harold G., and Thomas Hurka. *The Greenhouse Effect: Ethics & Climate Change.* Waterloo: Wilfrid Laurier University Press, 1993.

Crawford, Angus. "Child Marriages Blight Bangladesh." BBC News. http://www.bbc.co.uk/news/magazine-17779413 (accessed September 4, 2018).

Cross, Frank Moore. "The Priestly Houses of Early Israel." In *Canaanite Myth and Hebrew Epic: Essays in the History of the Religion of Israel*, 195–216. Cambridge, MA: Harvard University Press, 1973.

Crossan, John Dominic. *Jesus: A Revolutionary Biography.* San Francisco: Harper, 1994.

Daily, Gretchen C. *Nature's Services: Societal Dependence on Natural Ecosystems.* Washington, DC: Island, 1997.

Daly, Jonathan. *The Rise of Western Power: A Comparative History of Western Civilization.* London: Bloomsbury, 2013.

Darwin, Charles. *On the Origin of Species by Means of Natural Selection, Or, the Preservation of Favoured Races in the Struggle for Life.* London: John Murray, 1859.

De La Torre, Miguel A. *Doing Christian Ethics from the Margins.* Maryknoll: Orbis, 2004.

Deane-Drummond, Celia. *Christ and Evolution: Wonder and Wisdom.* Minneapolis: Fortress, 2009.

Deane-Drummond, Celia, and David Clough. *Creaturely Theology: On God, Humans and Other Animals.* London: SCM, 2009.

DeCrane, Susanne M. *Aquinas, Feminism, and the Common Good.* Washington, DC: Georgetown University Press, 2004.

Delio, Ilia, et al. *Care for Creation: A Franciscan Spirituality of the Earth.* Cincinnati: St. Anthony Messenger, 2007.

Derrida, Jacques, and Marie-Louise Mallet. *The Animal that therefore I Am.* New York: Fordham University Press, 2008.

Descartes, René. "Animals are Machines." In *Environmental Ethics: Divergence & Convergence*, edited by Susan J. Armstrong and Richard G. Botzler, 3rd ed., 274–78. New York: McGraw-Hill, 2003.

Devall, Bill, and George Sessions. *Deep Ecology: Living as if Nature Mattered.* Salt Lake City, UT: Peregrine Smith, 1985.

DeYoung, Curtiss Paul. *Living Faith: How Faith Inspires Social Justice.* Minneapolis: Fortress, 2007.

Diamond, Jared M. *Collapse: How Societies Choose to Fail Or Succeed.* New York: Penguin, 2005.

Dillard, Annie. *Pilgrim at Tinker Creek.* New York: Harper Perennial, 1998.

Donner, Simon. "Sea Level Rise and the Ongoing Battle of Tarawa." *EOS* 93, no. 17 (April 24, 2012) 169–70.

Dower, Nigel. "The Earth Charter and Global Ethics." *World Views: Environment, Culture, Religion* 8, no. 1 (2004) 15–28.

Droel, William L. *What is Social Justice?* Chicago: ACTA, 2011.

Duminuco, Vincent J. *The Jesuit Ratio Studiorum: 400th Anniversary Perspectives.* New York: Fordham University Press, 2000.

Dunlap, Thomas R. *Faith in Nature: Environmentalism as Religious Quest.* Seattle: University of Washington Press, 2004.

Bibliography

Durkheim, Emile. *The Division of Labour in Society*. London: Macmillan, 1984 (1893).
Durning, Alan Thein. *How Much is Enough? The Consumer Society and the Future of the Earth*. New York: Norton, 1992.
Eaton, Heather. "The Revolution of Evolution." *Worldviews: Global Religions, Culture & Ecology* 11, no. 1 (03, 2007) 6–31.
Edwards, Andres R. *The Sustainability Revolution: Portrait of a Paradigm Shift*. Gabriola, BC: New Society, 2005.
Eggemeier, Matthew T. "Ecology and Vision." *Worldviews: Global Religions, Culture & Ecology* 18, no. 1 (March, 2014) 54–76.
Ehrlich, Paul R., and Anne H. Ehrlich. *Extinction: The Causes and Consequences of the Disappearance of Species*. New York: Ballantine, 1983.
Eisen, Jessica. "Liberating Animal Law: Breaking Free from Human-use Typologies." *Animal Law* 17, no. 1 (2010) 59–76.
Elgin, Duane. *Promise Ahead: A Vision of Hope and Action for Humanity's Future*. New York: W. Morrow, 2000.
Elliott, James R., and Jeremy Pais. "Race, Class, and Hurricane Katrina: Social Differences in Human Responses to Disaster." *Social Science Research*. 35, no. 2 (2006) 295–321.
Elsbernd, Mary, and Reimund Bieringer. *When Love is Not enough: A Theo-Ethic of Justice*. Collegeville: Liturgical, 2002.
Engel, J. Ronald. "A Covenant Model of Global Ethics." *Worldviews: Environment, Culture, Religion* 8, no. 1 (2004) 29–46.
Evangelical Lutheran Church in America. "Caring for Creation: Vision, Hope, and Justice." In *This Sacred Earth: Religion, Nature, Environment*, edited by Roger S. Gottlieb, 2nd ed., 215–22. New York: Routledge, 1993.
Evans, Timothy, Margaret Whitehead, Finn Diderichsen, Abbas Bhuiya, and Meg Wirth, eds. *Challenging Inequities in Health from Ethics to Action*. New York: Oxford University Press, 2001.
Fagan, Brian M. *The Attacking Ocean: The Past, Present, and Future of Rising Sea Levels*. New York: Bloomsbury, 2013a.
———. "The Impending Deluge." *The New York Times*, May 31, 2013. https://www.nytimes.com/2013/06/01/opinion/global/brian-fagan-the-impending-deluge.html (accessed September 4, 2018).
Farmer, Paul. *Infections and Inequalities: The Modern Plagues*. Berkeley: University of California Press, 2001.
Fasching, Darrell J., and Dell deChant. *Comparative Religious Ethics: A Narrative Approach*. 1st ed. Oxford: Wiley-Blackwell, 2001.
Fasching, Darrell J., Dell deChant, and David M. Lantigua. *Comparative Religious Ethics: A Narrative Approach to Global Ethics*. 2nd ed. Oxford: Wiley-Blackwell, 2011.
Fiala, Nathan. "Measuring Sustainability: Why the Ecological Footprint is Bad Economics and Bad Environmental Science." *Ecological Economics Ecological Economics* 67, no. 4 (2008) 519–25.
Flanagan, Tara. "The Broken Body of God: Moving Beyond the Beauty Bias in Ecological Ethics." *Currents in Theology and Mission* 39, no. 2 (2012) 146–50.
Fleming, Melissa. "Climate Change could Become the Biggest Driver of Displacement: UNHCR Chief." UNHCR: The UN Refugee Agency. http://www.unhcr.org/4b2910239.html (accessed September 4, 2018).
Forell, George Wolfgang. *History of Christian Ethics: From the New Testament to Augustine*. Vol. 1. Minneapolis: Augsburg, 1979.

Bibliography

Fortune, Marie M. *Sexual Violence: The Unmentionable Sin*. New York: Pilgrim, 1983.

Francis, Pope. *Laudato Si', Encyclical Letter*. Vatican City, Italy: Libreria Editrice Vaticana, 2015. http://w2.vatican.va/content/francesco/en/encyclicals/documents/papa-francesco_20150524_enciclica-laudato-si.html (Accessed May 15, 2017).

Frayne, Bruce, et al. *Climate Change, Assets and Food Security in Southern African Cities*. New York: Earthscan, 2012.

French, William. "Ecology." In *The Blackwell Companion to Religious Ethics*, edited by William Schweiker, 469–76. Malden, MA: Blackwell, 2005.

———. "Beast-Machines and the Technocratic Reduction of Life." In *Good News for Animals? Christian Approaches to Animal Well-Being*, edited by Charles Robert Pinches and Jay B. McDaniel, 24–43. Maryknoll, NY: Orbis, 1993.

———. "Grace is Everywhere: Thomas Aquinas on Creation and Salvation." In *Creation and Salvation: A Mosaic of Selected Classic Christian Theologies*, edited by Ernst M. Conradie, vol. 1, 147–72. Zurich: LIT Verlag, 2012.

———. "Greening *Gaudium Et Spes*." In *Vatican II: Forty Years Later*, edited by William Madges, 196–207. Maryknoll: Orbis, 2006.

———. "Natural Law and Ecological Responsibility: Drawing on the Thomistic Tradition." *University of St. Thomas Law Journal* 5, no. 1 (Winter 2008) 12–36.

———. "On Knowing Oneself in an Age of Ecological Concern." In *Confronting the Climate Crisis: Catholic Theological Perspectives*, edited by Jame Schaefer, 145–76. Milwaukee: Marquette University Press, 2011.

———. "With Radical Amazement: Ecology and the Recovery of Creation." In *Without Nature? A New Condition for Theology*, edited by David Albertson and Cabell King, 54–79. New York: Fordham University Press, 2010.

———. "The World as God's Body: Theological Ethics and Panentheism." In *Broken and Whole: Essays on Religion and the Body*, edited by Maureen A. Tilley and Susan A. Ross, 135–44. Lanham: University Press of America, 1995.

Freud, Sigmund, and James Strachey. *The Future of an Illusion*. New York: Norton, 1975.

Friedman, Thomas L. *Hot, Flat, and Crowded: Why we Need a Green Revolution—and how it can Renew America*. New York: Farrar, Straus and Giroux, 2008.

Fromm, Harold. *The Nature of being Human: From Environmentalism to Consciousness*. Baltimore: Johns Hopkins University Press, 2009.

Galbraith, Kyle L. "Broken Bodies of God: The Christian Eucharist as a Locus for Ecological Reflection." *Worldviews: Global Religions, Culture & Ecology* 13, no. 3 (October 2009) 283–304.

Gallup, John Luke and Jeffrey D. Sachs. *The Economic Burden of Malaria*. Cambridge, MA: Center for International Development at Harvard University, 2000.

Garanzini, Michael J. "Faculty Convocation 2012." Loyola University Chicago. http://www.luc.edu/president/communications/facultyconvocation/archive/facultyconvocation2012/ (accessed September 25, 2013).

Gardiner, Stephen Mark. *A Perfect Moral Storm: The Ethical Tragedy of Climate Change*. New York: Oxford University Press, 2011.

Garvey, James. *The Ethics of Climate Change: Right and Wrong in a Warming World*. London: Continuum, 2008.

Gebara, Ivone. "Ecofeminism: An Ethics of Life." In *Ecofeminism & Globalization: Exploring Culture, Context, and Religion*, 163–76. Lanham: Rowman & Littlefield, 2003.

Bibliography

———. *Longing for Running Water: Ecofeminism and Liberation*. Minneapolis: Fortress, 1999.
———. *Out of the Depths: Women's Experience of Evil and Salvation*. Minneapolis: Fortress, 2002.
Geertz, Clifford. *The Interpretation of Cultures: Selected Essays*. New York: Basic, 1973.
Gemenne, François. "Climate-Induced Population Displacements in a 4°C+ World." *Philosophical Transactions of the Royal Society A: Mathematical, Physical & Engineering Sciences* 369, no. 1934 (2011) 182–95.
Gerrard, Michael B., and Gregory E. Wannier, eds. *Threatened Island Nations: Legal Implications of Rising Seas and a Changing Climate*. Cambridge: Cambridge University Press, 2013.
Gibson, William E. *Eco-Justice: The Unfinished Journey*. Albany: State University of New York Press, 2004.
Gillis, Justin. "A Climate Alarm, Too Muted for Some." *The New York Times*, September 9, 2013. https://www.nytimes.com/2013/09/10/science/a-climate-alarm-too-muted-for-some.html (accessed September 4, 2018).
———. "U.N. Climate Panel Endorses Ceiling on Global Emissions." *The New York Times*, September 27, 2013. https://www.nytimes.com/2013/09/28/science/global-climate-change-report.html (accessed September 4, 2018).
Goldman, Michael. *Privatizing Nature: Political Struggles for the Global Commons*. New Brunswick: Rutgers University Press, 1998.
Goodall, Jane. *Reason for Hope: A Spiritual Journey*. New York: Warner, 1999.
Gore, Albert. *Earth in the Balance: Ecology and the Human Spirit*. Boston: Houghton Mifflin, 1992.
———. *An Inconvenient Truth*. Directed by Davis Guggenheim. Hollywood: Paramount Classics, 2006.
———. *An Inconvenient Truth: The Planetary Emergency of Global Warming and What We Can Do about It*. New York: Rodale, 2006.
Gottlieb, Roger S. *A Greener Faith: Religious Environmentalism and our Planet's Future*. Oxford: Oxford University Press, 2006.
———. *This Sacred Earth: Religion, Nature, Environment*. 2nd ed. New York: Routledge, 2003.
Gould, Stephen Jay. "The Golden Rule—a Proper Scale for our Environmental Crisis." *Natural History* 99, no. 9 (1990) 24–30.
Grandin, Karl, ed. *The Nobel Prize 2007. Les Prix Nobel/Nobel Lectures*. Stockholm: The Nobel Foundation, 2008.
Grasso, Kenneth L., Gerard V. Bradley, and Robert P. Hunt, eds. *Catholicism, Liberalism, and Communitarianism: The Catholic Intellectual Tradition and the Moral Foundations of Democracy*. Lanham: Rowman & Littlefield, 1995.
Grim, John. "Time, History, Historians in Thomas Berry's Vision." The Thomas Berry Foundation. www.ThomasBerry.org (accessed October, 2014).
Grim, John A. *The Shaman: Patterns of Siberian and Ojibway Healing*. 2nd ed. Norman, OK: University of Oklahoma Press, 1987.
Grim, John A., ed. *Indigenous Traditions and Ecology: The Interbeing of Cosmology and Community*. Cambridge, MA: Harvard University Press, 2001.
Grim, John, and Mary Evelyn Tucker. *Ecology and Religion*. Washington, DC: Island, 2014.
Groody, Daniel G., and Gustavo Gutiérez, eds. *The Preferential Option for the Poor Beyond Theology*. Notre Dame: University of Notre Dame Press, 2013.

Bibliography

Grossmann, Iris, and M. Granger Morgan. "Tropical Cyclones, Climate Change, and Scientific Uncertainty: What do we Know, what does it Mean, and what should be done?" *Climatic Change* 108, no. 3 (2011) 543–79.

Gudorf, Christine E. "To make a Seamless Garment, use a Single Piece of Cloth." *Cross Currents* Winter (1984–85) 473–91.

Gustafson, James M. *Ethics from a Theocentric Perspective.* Vol. 2. Chicago: University of Chicago Press, 1984.

———. *Ethics from a Theocentric Perspective.* Vol. 1. Chicago: University of Chicago Press, 1983.

———. "Introduction." In *The Responsible Self: An Essay in Christian Moral Philosophy*, edited by H. Richard Niebuhr, 6–41. New York: Harper & Row, 1963.

Guterres, Antonio. "Millions Uprooted: Saving Refugees and the Displaced." *Foreign Affairs* 87, no. 5 (2008) 90–99.

Gutiérrez, Gustavo. *The God of Life.* Maryknoll: Orbis, 1991.

———. "Preferential Option for the Poor." In *Gustavo Gutiérrez: Essential Writings*, edited by James B. Nickoloff, 143–45. Maryknoll: Orbis, 1996.

———. *A Theology of Liberation: History, Politics, and Salvation.* Maryknoll: Orbis, 1973.

Hall, Douglas John. *The Steward: A Biblical Symbol Come of Age.* Grand Rapids: W. B. Eerdmans, 1990.

Hall, Douglas John. *Professing the Faith: Christian Theology in a North American Context.* Minneapolis: Fortress, 1993.

Hance, Jeremy. "A Key Mangrove Forest Faces Major Threat from a Coal Plant." *Yale Environment 360: Opinion, Analysis, Reporting and Debate*, October 29, 2013. https://e360.yale.edu/features/a_key_mangrove_forest_faces_major_threat_from_a_coal_plant (accessed September 4, 2018).

Haraway, Donna Jeanne. *When Species Meet.* Minneapolis: University of Minnesota Press, 2008.

Hargrove, Eugene C. *The Animal Rights, Environmental Ethics Debate: The Environmental Perspective.* Albany: State University of New York Press, 1992.

Häring, Bernhard. *God's Word and Man's Response.* New York: Paulist, 1963.

———. *The Law of Christ: Moral Theology for Priest and Laity.* Translated by Edwin G. Kaiser. Vol. 1. Paramus: Newman, 1961.

Harris, Paul G. and Steve Vanderheiden. *Ethics and Global Environmental Policy: Cosmopolitan Conceptions of Climate Change.* Cheltenham: Edward Elgar, 2011.

Harrison, Kathryn and Lisa McIntosh Sundstrom. *Global Commons, Domestic Decisions: The Comparative Politics of Climate Change.* Cambridge, MA: MIT Press, 2010.

Hauerwas, Stanley. *A Community of Character: Toward a Constructive Christian Social Ethic.* Notre Dame: University of Notre Dame Press, 1981.

Haught, John F. *Making Sense of Evolution: Darwin, God, and the Drama of Life.* Louisville: Westminster John Knox, 2010.

Hebblethwaite, Peter. "Liberation Theology and the Roman Catholic Church." In *The Cambridge Companion to Liberation Theology*, edited by Christopher Rowland, 179–98. Cambridge: Cambridge University Press, 1999.

Heer, James. *The President's Dilemma.* Film. Directed by James Heer. Oley, PA: Television Trust for the Environment; BBC World Service; Bullfrog Films, 2011.

Hessel, Dieter T. *After Nature's Revolt: Eco-Justice and Theology.* Minneapolis: Fortress, 1992.

Bibliography

Hessel, Dieter T., and Larry L. Rasmussen. *Earth Habitat: Eco-Injustice and the Church's Response*. Minneapolis: Fortress, 2001.

Hessel, Dieter T., and Rosemary Radford Ruether, eds. *Christianity and Ecology: Seeking the Well-being of Earth and Humans*. Cambridge, MA: Harvard University Press, 2000.

Heyd, Thomas, ed. *Recognizing the Autonomy of Nature: Theory and Practice*. New York: Columbia University Press, 2005.

Hiebert, Theodore. "The Human Vocation: Origins and Transformations in Christian Traditions." In *Christianity and Ecology: Seeking the Well-being of Earth and Humans*, edited by Dieter T. Hessel and Rosemary Radford Ruether, 135–54. Cambridge, MA: Harvard University Press, 2000.

———. *The Yahwist's Landscape: Nature and Religion in Early Israel*. New York: Oxford University Press, 1996.

Honeyman, Ryan. *The B Corp Handbook: How to Use Business as a Force for Good*. Oakland, CA: Berret-Koehler, 2014.

Hughes, J. Donald. *North American Indian Ecology*. 2nd ed. El Paso: Texas Western, 1996.

Hunt, James B. *Restless Fires: Young John Muir's Thousand-Mile Walk to the Gulf in 1867/68*. Macon, GA: Mercer University Press, 2012.

Huq, Saleemul, et al. "Sea-Level Rise and Bangladesh: A Preliminary Analysis." *Journal of Coastal Research* Spring, no. 14 (1995) 44–53.

Intergovernmental Panel on Climate Change. Working Group I. *Climate Change 2013: The Physical Science Basis: Contribution of Working Group I to the Fifth Assessment Report of the Intergovernmental Panel on Climate Change*. Cambridge: Cambridge University Press, 2013.

International Energy Agency. "Energy for all: Financing Access for the Poor." In *World Energy Outlook*. Oslo: International Energy Agency, 2011. https://www.iea.org/publications/freepublications/publication/weo2011_energy_for_all.pdf

Isasi-Díaz, Ada María. *Mujerista Theology: A Theology for the Twenty-First Century*. Maryknoll: Orbis, 1996.

Jasper, Scott. *Conflict and Cooperation in the Global Commons: A Comprehensive Approach for International Security*. Washington, DC: Georgetown University Press, 2012.

Jenkins, Willis. *Ecologies of Grace: Environmental Ethics and Christian Theology*. Oxford: Oxford University Press, 2008a.

———. "Global Ethics, Christian Theology, and the Challenge of Sustainability." *Worldviews: Global Religions, Culture & Ecology* 12, no. 2 (07, 2008b) 197–217.

Jetley, Surinder. "Impact of Male Migration on Rural Females." *Economic and Political Weekly* 22, no. 44 (October 31, 1987) 47–53.

John XXIII, Pope. "*Pacem in Terris*: Peace on Earth (1963)." In *Catholic Social Thought: The Documentary Heritage*, edited by David J. O'Brien and Thomas A. Shannon, 129–62. Maryknoll: Orbis, 2005.

John Paul II, Pope. *Encyclical Letter Centesimus Annus of the Supreme Pontiff John Paul II: On the Hundredth Anniversary of Rerum Novarum*. Boston: St. Paul & Media, 1991.

———. "Laborem Exercens: On Human Work (1981)." In *Catholic Social Thought: The Documentary Heritage*. Edited by David J. O'Brien and Thomas A. Shannon, 350–92. Maryknoll: Orbis, 2005.

———. "Sollicitudo Rei Socialis: On Social Concern *(1987)*." In *Catholic Social Thought: The Documentary Heritage*, edited by David J. O'Brien and Thomas A. Shannon, 393–436. Mary Knoll: Orbis, 2005.

Bibliography

Johnson, Elizabeth A. *Ask the Beasts: Darwin and the God of Love*. London: Bloomsbury, 2014.

———. "Losing and Finding Creation in the Christian Tradition." In *Christianity and Ecology: Seeking the Well-being of Earth and Humans*, edited by Dieter Hessel and Rosemary Radford Reuther, 3–22. Cambridge, MA: Harvard University Press, 2000.

Jonas, Hans. *The Imperative of Responsibility: In Search of an Ethics for the Technological Age*. Chicago: University of Chicago Press, 1984.

Jones, Nicola. "Rising Waters: How Fast and how Far Will Sea Levels Rise?" *Yale Environment 360: Opinion, Analysis, Reporting and Debate*, October 21, 2013. https://e360.yale.edu/features/rising_waters_how_fast_and_how_far_will_sea_levels_rise (accessed September 4, 2018).

Juergensmeyer, Mark, ed. *Religion in Global Civil Society*. Oxford: Oxford University Press, 2005.

Jung, Patricia Beattie, and Aana Marie Vigen, eds. *God, Science, Sex, Gender: An Interdisciplinary Approach to Christian Ethics*. Champaign: University of Illinois Press, 2010.

Kahan, Dan M., et al. "The Polarizing Impact of Science Literacy and Numeracy on Perceived Climate Change Risks." *Nature Climate Change* 2, no. 10 (October 2012) 732–35.

Kant, Immanuel. *Groundwork of the Metaphysics of Morals*. Translated by Mary J. Gregor. Cambridge: Cambridge University Press, 1998.

Keevey, Thomas. "Thomas Berry, C.P.: The Passionist Heritage in the Great Work." University of North Carolina at Chapel Hill, paper presented at the "Colloquium on Thomas Berry's Work: Development, Difference, Importance, Applications" sponsored by the Center for Ecozoic Societies and Carolina Seminars of the University of North Carolina at Chapel Hill, 28–30 May 2014.

Kirkpatrick, Martha, ed. *Women's Sexual Experience: Exploration of the Dark Continent*. New York: Plenum, 1982.

Klare, Michael T. *Rising Powers, Shrinking Planet: The New Geopolitics of Energy*. New York: Metropolitan, 2008.

Klinenberg, Eric. *Heat Wave: A Social Autopsy of Disaster in Chicago*. Chicago: University of Chicago Press, 2002.

Klinkenborg, Verlyn. "Animal 'Personhood': Muddled Alternative to Real Protection." *Yale Environment 360: Opinion, Analysis, Reporting and Debate*, January 30, 2014. https://e360.yale.edu/features/animal_personhood_muddled_alternative_to_real_protection (accessed September 4, 2018).

Klyza, Christopher McGrory. "Do Trees have Rights? Rights, Nature, and Conceptual Change." *Southeastern Political Review* 22, no. 3 (1994) 427–44.

Kristof, Nicholas D., and Sheryl WuDunn. *Half the Sky: Turning Oppression into Opportunity for Women Worldwide*. New York: Alfred A. Knopf, 2009.

Kuhn, Thomas S. *The Copernican Revolution: Planetary Astronomy in the Development of Western Thought*. Cambridge, MA: Harvard University Press, 1957.

Kuhn, Thomas S. *The Structure of Scientific Revolutions*. 1st ed. Chicago: University of Chicago Press, 1962.

Kwok, Roberta. "'Greening of Christianity'? Not Yet." *Conservation this Week: The Source for Environmental Intelligence*, July 26, 2013. http://www.conservationmagazine.org/2013/07/greening-of-christianity-not-yet/ (accessed September 4, 2018).

BIBLIOGRAPHY

Lakoff, George. *Moral Politics: What Conservatives Know that Liberals Don't*. Chicago: University of Chicago Press, 1996.

Lakoff, George and Mark Johnson. *Metaphors we Live By*. Chicago: University of Chicago Press, 1980.

Lane, John. "Lake Conestee." In *A Voice for Earth: American Writers Respond to the Earth Charter*, edited by Peter Blaze Corcoran et al., 61–67. Athens: University of Georgia Press, 2008.

Lang, Chang, and Darryn W. Waugh. "Impact of Climate Change on the Frequency of Northern Hemisphere Summer Cyclones." *Journal of Geophysical Research* 116, no. D04103 (2011) 1–12.

Laszlo, Ervin, and Allan Combs. *Thomas Berry, Dreamer of the Earth: The Spiritual Ecology of the Father of Environmentalism*. Rochester: Inner Traditions, 2011.

Le Treut, H., et al. "2007: Historical Overview of Climate Change." In *Climate Change 2007: The Physical Science Basis. Contribution of Working Group I to the Fourth Assessment Report of the Intergovernmental Panel on Climate Change*, edited by S. Solomon, et al. Cambridge: Cambridge University Press, 2007. https://www.ipcc.ch/publications_and_data/ar4/wg1/en/ch1.html

Leakey, Richard E., and Roger Lewin. *The Sixth Extinction: Patterns of Life and the Future of Humankind*. New York: Anchor, 1996.

Lebacqz, Karen. *Justice in an Unjust World: Foundations for a Christian Approach to Justice*. Minneapolis: Augsburg, 1987.

———. *Six Theories of Justice: Perspectives from Philosophical and Theological Ethics*. Minneapolis: Augsburg, 1986.

LeBlanc, Jill. "Eco-Thomism." *Environmental Ethics* 21, no. 3 (1999) 293–306.

Ledoux, Arthur O. "A Green Augustine: On Learning to Love Nature Well." *Theology and Science* 3, no. 3 (2005) 331–44.

Lemonick, Mich. "As Effects of Warming Grow, U.N. Report is Quickly Dated." *Yale Environment 360: Opinion, Analysis, Reporting and Debate*, February 12, 2009. https://e360.yale.edu/features/as_effects_of_warming_grow_un_report_is_quickly_dated (accessed September 4, 2018).

Leopold, Aldo. *A Sand County Almanac, and Sketches here and There*. New York: Oxford University Press, 1949.

Lerner, Michael. *Jewish Renewal: A Path to Healing and Transformation*. New York: Putnam, 1995.

Lin, Xiaodong. *Gender, Modernity and Male Migrant Workers in China: Becoming a 'Modern' Man*. New York: Routledge, 2013.

Linzey, Andrew. *Why Animal Suffering Matters: Philosophy, Theology, and Practical Ethics*. Oxford: Oxford University Press, 2009.

———. *Animal Theology*. Urbana: University of Illinois Press, 1995.

Linzey, Andrew, and Dan Cohn-Sherbok. *After Noah: Animals and the Liberation of Theology*. Herndon: Mowbray, 1997.

Linzey, Andrew, and Dorothy Yamamoto. *Animals on the Agenda: Questions about Animals for Theology and Ethics*. Urbana: University of Illinois Press, 1998.

Lodge, David M., and Christopher Hamlin, eds. *Religion and the New Ecology: Environmental Responsibility in a World in Flux*. Notre Dame: University of Notre Dame Press, 2006.

Logan, John R. *The Impact of Katrina: Race and Class in Storm-Damaged Neighborhoods*. Providence: Brown University, 2006.

Bibliography

Lorentzen, Lois Ann. "Indigenous Feet: Ecofeminism, Globalization, and the Case of Chiapas." In *Ecofeminism and Globalization: Exploring Culture, Context, and Religion*, edited by Heather Eaton and Lois Ann Lorentzen, 57–72. Oxford: Rowman & Littlefield, 2003.

Lovejoy, Arthur O. *The Great Chain of Being: A Study of the History of an Idea*. Cambridge, MA: Harvard University Press, 1936.

Loy, David R. "The Religion of the Market." *Journal of the American Academy of Religion* 65, no. 2 (1997) 275–90.

Ludwig, Robert A. *Reconstructing Catholicism: for a New Generation*. Eugene, OR: Wipf & Stock, 1995.

Lurz, Robert W., ed. *The Philosophy of Animal Minds*. Cambridge: Cambridge University Press, 2009.

Lushwala, Arkan. *The Time of the Black Jaguar: An Offering of Indigenous Wisdom for the Continuity of Life on Earth*. San Bernardino: Createspace, 2013.

Luterbacher, Urs, and Detlef F. Sprinz, eds. *International Relations and Global Climate Change*. Cambridge, MA: MIT Press, 2001.

Lutheran Health Care Bangladesh (LHCB-USA). "Improving Lives through Health Care and Education." Lutheran Health Care Bangladesh. http://lhcb.org (accessed November 6, 2013).

Macpherson, C. B. *The Political Theory of Possessive Individualism: Hobbes to Locke*. Oxford: Clarendon, 1962.

Mahoney, John. *The Making of Moral Theology: A Study of the Roman Catholic Tradition*. Oxford: Oxford University Press, 1987.

Malinowski, Bronislaw. *The Ethnography of Malinowski: The Trobriand Islands*. London: Routledge, 1979.

Mander, Jerry. *In the Absence of the Sacred: The Failure of Technology and the Survival of the Indian Nations*. San Francisco: Sierra Club, 1991.

Maplecroft (Firm). *Climate Change and Environmental Risk Atlas*. 5th ed. Bath: Maplecroft, 2013.

———. "Who we Are." Maplecroft. www.maplecroft.com (accessed May 24, 2013).

Marshall, Joseph. *To You we Shall Return: Lessons about our Planet from the Lakota*. New York: Sterling Ethos, 2010.

Marx, Karl. "Toward A Critique of Hegel's *Philosophy of Right*: Introduction." In *Karl Marx: Selected Writings*, edited by Lawrence H. Simon, translated by Loyd D. Easton and Kurt H. Guddat, 27–39. Indianapolis: Hackett, 1994.

Massaro, Thomas, SJ. *Living Justice: Catholic Social Teaching in Action*. Lanham: Rowman & Littlefield, 2012.

Mastaler, James S. "A Case Study on Climate Change and its Effects on the Global Poor." *Worldviews: Global Religions, Culture, and Ecology* 15, no. 1 (2011) 65–87.

———. "The Role of Christian Ethics, Religious Leaders, and People of Faith at a Time of Ecological and Climate Crisis." New Theology Review 26, no. 2 (2014) 43–48.

McFague, Sallie. *The Body of God: An Ecological Theology*. Minneapolis: Fortress, 1993.

———. *Life Abundant: Rethinking Theology and Economy for a Planet in Peril*. Minneapolis: Fortress, 2001.

———. *Metaphorical Theology: Models of God in Religious Language*. Philadelphia: Fortress, 1982.

———. *Models of God: Theology for an Ecological, Nuclear Age*. Philadelphia: Fortress, 1987.

Bibliography

———. *A New Climate for Theology: God, the World, and Global Warming.* Minneapolis: Fortress, 2008.

McGrath, Alister E. *Christian Theology: An Introduction.* 6th ed. Oxford: Wiley Blackwell, 2017.

McIntosh, Alastair. "A New Climate for Theology." *Journal for the Study of Religion, Nature & Culture* 5, no. 3 (September 2011) 384–86.

McKibben, Bill. *The Age of Missing Information.* New York: Random House, 1992.

———. *The Comforting Whirlwind: God, Job, and the Scale of Creation.* Grand Rapids: W. B. Eerdmans, 1994.

———. *Earth: Making a Life on a Tough New Planet.* New York: Times, 2010.

Merchant, Carolyn. *The Death of Nature: Women, Ecology, and the Scientific Revolution.* New York: Harper & Row, 1989.

———. *Reinventing Eden: The Fate of Nature in Western Culture.* 2nd ed. New York: Routledge, 2013.

Mesoudi, Alex. *Cultural Evolution: How Darwinian Theory can Explain Human Culture and Synthesize the Social Sciences.* Chicago: University of Chicago Press, 2011.

Midgley, Mary. *Animals and Why they Matter.* Athens, GA: University of Georgia Press, 1984.

———. *Beast and Man: The Roots of Human Nature.* Revised ed. London: Routledge Classics, 1995.

Miles, Margaret R. *Augustine on the Body.* Eugene, OR: Wipf & Stock, 2009.

Miller, Peter, and Laura Westra. *Just Ecological Integrity: The Ethics of Maintaining Planetary Life.* Lanham: Rowman & Littlefield, 2002.

Milun, Kathryn. *The Political Uncommons: The Cross-Cultural Logic of the Global Commons.* Farnham: Ashgate, 2011.

Min, Seung-Ki, et al. "Human Contribution to More-Intense Precipitation Extremes." *Nature* 470, no. 7334 (2011) 378.

Mirza, M. Monirul Qader. "Global Warming and Changes in the Probability of Occurrence of Floods in Bangladesh and Implications." *Global Environmental Change* (2002) 127–38.

Moe-Lobeda, Cynthia D. *Resisting Structural Evil: Love as Ecological and Economic Vocation.* Minneapolis: Fortress, 2013.

Mohamed Salih, M. A., ed. *Inducing Food Insecurity: Perspectives on Food Policies in Eastern and Southern Africa.* Uppsala: Nordiska Afrikainstitutet, 1994.

Moltmann, Jürgen. *The Spirit of Life: A Universal Affirmation.* Minneapolis: Fortress, 1992.

Mooney, Chris. "The Science of Why we Don't Believe Science: How our Brains Fool Us on Climate, Creationism, and the Vaccine-Autism Link." *Mother Jones* 36, no. 3 (May/June 2011) 40–45.

Mooney, Harold A., and Paul R. Ehrlich. "Ecosystem Services: A Fragmentary History." In *Nature's Services: Societal Dependence on Natural Ecosystems*, edited by Gretchen C. Daily, 11–19. Washington, DC: Island, 1997.

Moore, Kathleen D. "The Ethics of Adaptation to Global Warming." Center for Humans and Nature. http://www.humansandnature.org/earth-ethic-kathleen-dean-moore-response-81.php (accessed September 4, 2018).

Moore, Stephen, and Laurel Kearns. *Divinanimality: Animal Theory, Creaturely Theology.* New York: Fordham University Press, 2014.

Mora, Camilo, et al. "The Projected Timing of Climate Departure from Recent Variability." *Nature* 502, no. 7470 (October 10, 2013) 183–87.

Bibliography

Muir, John, and William Frederic Badè. *A Thousand-Mile Walk to the Gulf.* Boston: Houghton Mifflin, 1916.

Murdy, William H. "Anthropocentrism: A Modern Version." *Science* 187, no. 4182 (1975) 1168–1172.

Naess, Arne. *Ecology, Community, and Lifestyle: Outline of an Ecosophy.* Translated by David Rothenberg. Cambridge: Cambridge University Press, 1989.

Nanda, Ved P., ed. *Climate Change and Environmental Ethics.* New Brunswick: Transaction, 2011.

Nash, James A. *Loving Nature: Ecological Integrity and Christian Responsibility.* Nashville: Abingdon, 1991.

Nash, Roderick F. *The Rights of Nature: A History of Environmental Ethics.* Madison: University of Wisconsin Press, 1989.

Navarro, Mireya. "Weighing Sea Barriers as Protection for New York." *The New York Times*, November 7, 2012. https://www.nytimes.com/2012/11/08/nyregion/after-hurricane-sandy-debating-costly-sea-barriers-in-new-york-area.html (accessed September 4, 2018).

Nayamweru, Celia. "Women and Sacred Groves in Coastal Kenya: A Contribution to the Ecofeminist Debate." In *Ecofeminism and Globalization: Exploring Culture, Context, and Religion*, edited by Heather Eaton and Lois Ann Lorentzen, 41–56. Oxford: Rowman & Littlefield, 2003.

Nelson, James B. *Body Theology.* Louisville: Westminster/John Knox, 1992.

New Jersey Department of Community Affairs. *Community Development Block Grant Disaster Recovery: Action Plan.* Washington, DC: US Department of Housing and Urban Development, 2013.

Newsom, Carol A., and Sharon H. Ringe. *The Women's Bible Commentary.* Expanded ed. Louisville: Westminster/John Knox, 1998.

Nickel, James W., and Eduardo Viola. "Integrating Environmentalism and Human Rights." *Environmental Ethics* 16, no. 3 (1994) 265–73.

Niebuhr, H. Richard. *The Responsible Self: An Essay in Christian Moral Philosophy.* New York: Harper & Row, 1963.

Northcott, Michael S. *A Political Theology of Climate Change.* Grand Rapids: William B. Eerdmans, 2013.

———. *A Moral Climate: The Ethics of Global Warming.* Maryknoll: Orbis, 2007.

Swimme, Brian, Mary Evelyn Tucker, and John Grim. *Journey of the Universe.* DVD. Directed by Patsy Northcutt and David Kennard. California: Northcutt Productions, InCA Productions, KTCS Seattle, KQED—PBS, 2011.

Nobel Foundation. "The Nobel Peace Prize 2007." The Nobel Foundation. http://www.nobelprize.org/nobel_prizes/peace/laureates/2007/gore-bio.html (accessed September 4, 2018).

Norton, Bryan G. "Environmental Ethics and Nonhuman Rights." *Environmental Ethics* 4, no. 1 (1982) 17–36.

Núñez, Lautaro A., and Juan de Dios Vial Larraín. *El Patrimonio Arqueológico Chileno: Reflexiones Sobre El Futuro Del Pasado.* Santiago de Chile: Instituto de Chile, Academia Chilena de Ciencias Sociales, 1986.

Nussbaum, Martha Craven. *Creating Capabilities: The Human Development Approach.* Cambridge, MA: Belknap, 2011.

———. *Women and Human Development: The Capabilities Approach.* Cambridge: Cambridge University Press, 2000.

Bibliography

O'Brien, David J., and Thomas A. Shannon, eds. *Catholic Social Thought: The Documentary Heritage*. Maryknoll: Orbis, 2005.

O'Brien, Kevin J. *An Ethics of Biodiversity: Christianity, Ecology, and the Variety of Life*. Washington, DC: Georgetown University Press, 2010.

O'Connor, Thomas S. "'We are Part of Nature': Indigenous Peoples' Rights as a Basis for Environmental Protection in the Amazon Basin." *Colorado Journal of International Environmental Law & Policy* 5, no. 1 (Winter 1994) 193–211.

Oelschlaeger, Max. *Caring for Creation: An Ecumenical Approach to the Environmental Crisis*. New Haven: Yale University Press, 1994a.

———. *The Idea of Wilderness: From Prehistory to the Age of Ecology*. New Haven: Yale University Press, 1991.

Oelschlaeger, Max. *Caring for Creation: An Ecumenical Approach to the Environmental Crisis*. New Haven: Yale University Press, 1994b.

Olson, Dennis T. "Untying the Knot? Masculinity, Violence, and the Creation-Fall Story of Genesis 2–4." In *Engaging the Bible in a Gendered World: An Introduction to Feminist Biblical Interpretation in Honor of Katharine Doob Sakenfeld*, edited by Linda Day and Carolyn Pressler, 73–88. Louisville: Westminster John Knox, 2006.

Pachauri, R. K., and A. Reisinger, eds. *Contribution of Working Groups I, II and III to the Fourth Assessment Report of the Intergovernmental Panel on Climate Change*. Geneva, Switzerland: IPCC, 2007.

Page, L. Kristen. "Global Climate Change: Implications for Global Health." In *Christians, the Care of Creation & Global Climate Change*, edited by Lindy Scott, 24–35. Eugene, OR: Pickwick, 2008.

Paik, Eugene. "Left Out of Federal Sandy Relief, Owners of Second Homes Hope for Help." *The Star-Ledger*, May 19, 2013. https://www.nj.com/ocean/index.ssf/2013/05/left_out_of_federal_sandy_relief_owners_of_second_homes_hope_for_help.html (accessed September 4, 2018).

Pall, Pardeep, et al. "Anthropogenic Greenhouse Gas Contribution to Flood Risk in England and Wales in Autumn 2000." *Nature* 470, no. 7334 (2011) 382–85.

Parry, M. L., et al, eds. *Contribution of Working Group II to the Fourth Assessment Report of the Intergovernmental Panel on Climate Change, 2007*. Cambridge: Cambridge University Press, 2007.

Paul VI, Pope. "Octogesima Adveniens: A Call to Action on the Eightieth Anniversary of Rerum Novarum (1971)." In *Catholic Social Thought: The Documentary Heritage*, edited by David J. O'Brien and Thomas A. Shannon, 263–86. Maryknoll: Orbis, 2005.

Pfeffer W. T., et al. "Kinematic Constraints on Glacier Contributions to 21st-Century Sea-Level Rise." *Science* 321, no. 5894 (2008) 1340–1343.

Pinches, Charles Robert, and Jay B. McDaniel, eds. *Good News for Animals? Christian Approaches to Animal Well-Being*. Maryknoll: Orbis, 1993.

Pirard P, Vandentorren S, et al. "Summary of the Mortality Impact Assessment of the 2003 Heat Wave in France." *Euro Surveillance: Bulletin Européen Sur Les Maladies Transmissibles = European Communicable Disease Bulletin* 10, no. 7 (2005) 153–56.

Pixley, Jorge V., and Clodovis Boff. *The Bible, the Church, and the Poor*. Maryknoll: Orbis, 1989.

Pogge, Thomas. *World Poverty and Human Rights*. 2nd ed. Cambridge: Polity, 2008.

Pontifical Academy of Sciences. *Fate of Mountain Glaciers in the Anthropocene*. Vatican: Pontifical Academy of Sciences, May 11, 2011.

Bibliography

Pontificium Consilium de Iustitia et Pace. *Compendium of the Social Doctrine of the Church*. Cittá del Vaticano; Washington, DC: Libreria Editrice Vaticana; United States Conference of Catholic Bishops, 2004.

Porter, Christopher A. "The Religion of Consumption and Christian Neighbor Love." PhD diss., Loyola University Chicago, 2013.

Porter, Jean. *The Recovery of Virtue: The Relevance of Aquinas for Christian Ethics*. Louisville: Westminster/John Knox, 1990.

Povilitis, Anthony J. "On Assigning Rights to Animals and Nature." *Environmental Ethics* 2, no. 1 (1980) 67–71.

Pyper, Julia. "Drama Unfolds on Islands Facing a Watery End." *Environment & Energy*, March 18, 2013. https://www.eenews.net/special_reports/islands/stories/1059977973 (accessed September 4, 2018).

Radicella, Lucas. "Protecting Pachamama: Bolivia's New Environmental Law." *The Argentina Independent*, November 21, 2012. http://www.argentinaindependent.com/socialissues/environment/protecting-pachamama-bolivias-new-environmental-law/ (accessed September 4, 2018).

Raines, John C., ed. *Marx on Religion*. Philadelphia: Temple University Press, 2002.

Rajan, S. Irudaya, ed. *India Migration Report 2011: Migration, Identity and Conflict*. New Delhi: Routledge India, 2011.

Rasmussen, Larry L. *Earth Community, Earth Ethics*. Maryknoll: Orbis, 1996.

———. *Earth-Honoring Faith: Religious Ethics in a New Key*. New York: Oxford University Press, 2013.

———. "Lutheran Sacramental Imagination." Journal of Lutheran Ethics. http://www.elca.org/JLE/Articles/42 (accessed April, 2014).

Raworth, Kate. *A Safe and Just Space for Humanity: Can we Live within the Doughnut?* Oxford: Oxfam GB for Oxfam International, 2012.

Raymond, Christopher M., et al. "Ecosystem Services and Beyond: Using Multiple Metaphors to Understand Human-Environment Relationships." *BioScience* 63, no. 7 (2013) 536–46.

Redfearn, Jennifer, and Tim Metzger, dirs. *Sun Come Up*. DVD. New York: Sun Come Up, LLC, 2011.

Regan, Tom. *The Case for Animal Rights*. Berkeley: University of California Press, 1983.

Reill, Peter Hanns. *Vitalizing Nature in the Enlightenment*. Berkeley: University of California Press, 2005.

Reiner, Robert C., Jr., et al. "Highly Localized Sensitivity to Climate Forcing Drives Endemic Cholera in a Megacity." *Proceedings of the National Academy of Sciences of the United States of America* 109, no. 6 (2012) 2033–2036.

Rejón, Francisco Moreno. "Fundamental Moral Theory in the Theology of Liberation." In *Mysterium Liberationis: Fundamental Concepts of Liberation Theology*, edited by Ignacio Ellacuría and Jon Sobrino, 210–21. New York: Orbis, 1993.

Rempel, Henry, and Richard A. Lobdell. "The Role of Urban-to-rural Remittances in Rural Development." *Journal of Development Studies* 14, no. 3 (1978) 324–41.

Renaud, Fabrice, et al. *Control, Adapt Or Flee: How to Face Environmental Migration?* Bonn: United Nations University—Institute for Environment and Human Security, 2007.

Reuveny, Rafael. "Climate Change-Induced Migration and Violent Conflict." *Political Geography* 26, no. 6 (2007) 656–73.

Bibliography

Revkin, Andrew C. "Ecuador Constitution Grants Rights to Nature." *The New York Times*, September 29, 2008. https://dotearth.blogs.nytimes.com/2008/09/29/ecuador-constitution-grants-nature-rights/ (accessed September 4, 2018).

Richardson, Bob. "Biography of Annie Dillard." AnnieDillard.com. http://www.anniedillard.com/biography-by-bob-richardson.html (accessed September 4, 2018).

Ricoeur, Paul. *The Philosophy of Paul Ricoeur: An Anthology of His Work*. Edited by Charles E. Reagan and David Stewart. Boston: Beacon, 1978.

———. *Freud and Philosophy: An Essay on Interpretation*. New Haven: Yale University Press, 1970.

———. *The Symbolism of Evil*. Translated by Emerson Buchanan. New York: Harper & Row, 1967.

Rifkin, Jeremy. *The Zero Marginal Cost Society: The Internet of Things, the Collaborative Commons, and the Eclipse of Capitalism*. New York: Palgrave Macmillan, 2014.

Rockefeller, Steven C. "Crafting Principles for the Earth Charter." In *A Voice for Earth: American Writers Respond to the Earth Charter*, edited by Peter Blaze Corcoran, James Wohlpart and Brandon P. Hollingshead, 3–23. Athens, GA: University of Georgia Press, 2008.

———. "Global Ethics, International Law, and the Earth Charter." *Earth Ethics* 7, no. 3 (Spring/Summer 1996) 1–7.

Rockström, Johan, et al. "A Safe Operating Space for Humanity." *Nature* 461, no. 7263 (2009a) 472–75.

———. "Planetary Boundaries: Exploring the Safe Operating Space for Humanity." *Ecology and Society* 14, no. 2 (2009b) 1–33.

Rolston III, Holmes, "Challenges in Environmental Ethics." In *Environmental Philosophy: From Animal Rights to Radical Ecology*, edited by Michael E. Zimmerman et al., 2nd ed., 124–44. Upper Saddle River, NJ: Prentice Hall, 1998.

———. *Conserving Natural Value*. New York: Columbia University Press, 1994

———. "Does Nature Need to be Redeemed?" *Zygon: Journal of Religion & Science* 29, no. 2 (1994) 205–29.

———. *Environmental Ethics: Duties to and Values in the Natural World*. Philadelphia: Temple University Press, 1988.

———. "Ethical Responsibilities Toward Wildlife." *Journal of the American Veterinary Medical Association* 200, no. 5 (1992) 615–22.

———. "Feeding People versus Saving Nature?" In *World Hunger and Morality*, edited by William Aiken and Hugh LaFollette, 248–67. Englewood Cliffs, NJ: Prentice Hall, 1996.

———. "Kenosis and Nature." In *The Work of Love: Creation as Kenosis*, edited by John Polkinghorne, 43–65. Grand Rapids: W. B. Eerdmans, 2001.

———. "Loving Nature: Christian Environmental Ethics." In *Love and Christian Ethics: Tradition, Theory, and Society*, edited by Frederick V. Simmons and Brian C. Sorrells, 313–31. Washington: Georgetown University Press, 2016.

———. *A New Environmental Ethics: The Next Millennium for Life on Earth*. New York: Routledge, 2011.

———. *Philosophy Gone Wild: Environmental Ethics*. Buffalo: Prometheus, 1989.

———. *Science & Religion: A Critical Survey*. Philadelphia: Templeton Foundation, 2006 (1987).

Ross, Susan A. *Anthropology: Seeking Light and Beauty*. Collegeville: Liturgical, 2012.

Bibliography

———. *Extravagant Affections: A Feminist Sacramental Theology*. New York: Continuum, 1998.
Roy, Eleanor Ainge. "New Zealand River Granted Same Legal Rights as Human Being." *The Guardian*. March 16, 2017. https://www.theguardian.com/world/2017/mar/16/new-zealand-river-granted-same-legal-rights-as-human-being (accessed September 4, 2018).
Rowntree, B. Seebohm. *Poverty: A Study of Town Life*. New York: Policy, 1901.
Ruether, Rosemary Radford. *Gaia & God: An Ecofeminist Theology of Earth Healing*. San Francisco: Harper Collins, 1992.
Ruether, Rosemary Radford. *Integrating Ecofeminism, Globalization, and World Religions*. Lanham: Rowman & Littlefield, 2005.
Russell, Diane E. H. *Rape in Marriage*. New York: Macmillan, 1983.
Sachs, Jeffrey D. *The Price of Civilization: Reawakening American Virtue and Prosperity*. New York: Random House, 2011.
———. *The End of Poverty: Economic Possibilities for our Time*. New York: Penguin, 2005.
Safi, Michael. "Ganges and Yamuna rivers granted same legal rights as human beings." *The Guardian*. March 21, 2017. https://www.theguardian.com/world/2017/mar/21/ganges-and-yamuna-rivers-granted-same-legal-rights-as-human-beings (accessed September 4, 2018).
Sagan, Carl. *Pale Blue Dot: A Vision of the Human Future in Space*. New York: Random House, 1994.
Sagan, Carl, et al. "An Open Letter to the Religious Community." In *Ecology and Religion: Scientists Speak*, edited by John E. Carroll and Keith Warner, ii–vi. Quincy, IL: Franciscan, 1998.
Santmire, H. Paul. *The Travail of Nature: The Ambiguous Ecological Promise of Christian Theology*. Philadelphia: Fortress, 1985.
Sarris, Alexander, and Jamie Morrison. *Food Security in Africa: Market and Trade Policy for Staple Foods in Eastern and Southern Africa*. Northampton: Edward Elgar, 2010.
Saunders, Doug. *Arrival City: How the Largest Migration in History is Reshaping our World*. New York: Pantheon, 2010.
Schaefer, Jame. *Theological Foundations for Environmental Ethics: Reconstructing Patristic & Medieval Concepts*. Washington, DC: Georgetown University Press, 2009.
Scharen, Christian Batalden, and Aana Marie Vigen. *Ethnography as Christian Theology and Ethics*. London: Continuum, 2011.
Scharper, Stephen B. *Redeeming the Time: A Political Theology of the Environment*. New York: Continuum, 1997.
Scheid, Daniel P. "Thomas Aquinas, the Cosmic Common Good, & Climate Change." In *Confronting the Climate Crisis: Catholic Theological Perspectives*, edited by Jame Schaefer, 125–44. Milwaukee: Marquette University Press, 2011.
Schneider, S. H., et al. "2007: Assessing Key Vulnerabilities and the Risk from Climate Change." In *Climate Change 2007: Impacts, Adaptation and Vulnerability. Contribution of Working Group II to the Fourth Assessment Report of the Intergovernmental Panel on Climate Change*, edited by M. L. Parry et al, 779–810. Cambridge: Cambridge University Press, 2007.
Schuck, Michael J. *That they be One: The Social Teaching of the Papal Encyclicals 1740–1989*. Washington, DC: Georgetown University Press, 1991.
Schut, Michael. *Money & Faith: The Search for Enough*. Denver, CO: Morehouse Education Resources, 2008.

Bibliography

Sciglitano, Anthony C. "Hans Urs Von Balthasar & Deep Ecology: Toward a Doxological Ecology." In *Confronting the Climate Crisis: Catholic Theological Perspectives*, edited by Jame Schaefer, 277–300. Milwaukee: Marquette University Press, 2011.

Scully, Edgar. "The Place of the State in Society According to Thomas Aquinas." *The Thomist* 45, no. 3 (1981) 407–29.

Second Vatican Council. "Gaudium Et Spes: Pastoral Constitution on the Church in the Modern World (1965)." In *Catholic Social Thought: The Documentary Heritage*, edited by David J. O'Brien and Thomas A. Shannon, 164–237. Maryknoll: Orbis, 2005.

Sen, Amartya. *The Idea of Justice*. Cambridge: Belknap, 2009.

———. "Missing Women: Social Inequality Outweighs Women's Survival Advantage in Asia and North Africa." *British Medical Journal* 304, no. 6827 (1992) 587–88.

Shepherd, Andrew, et al. *The Geography of Poverty, Disasters and Climate Extremes in 2030*. London: Overseas Development Institute, 2013.

Shubin, Neil. *The Universe Within: Discovering the Common History of Rocks, Planets, and People*. New York: Pantheon, 2013.

Shweder, Richard A., and Byron Good, eds. *Clifford Geertz by His Colleagues*. Chicago: University of Chicago Press, 2005.

Sideris, Lisa. "Evolving Environmentalism: The Role of Ecotheology in Creation/Evolution Controversies." *Worldviews: Global Religions, Culture, and Ecology* 11, no. 1 (2007) 58–82.

Sigmund, Paul E., ed. *St. Thomas Aquinas on Politics and Ethics*. New York: W. W. Norton & Company, 1987.

Singer, Peter. *Animal Liberation: A New Ethics for our Treatment of Animals*. New York: Harper, 1975.

Smith, Adam. *An Inquiry into the Nature and Causes of the Wealth of Nations*. Oxford: Oxford University Press, 1998 (1779).

Society of Jesus. "Healing a Broken World: Task Force on Ecology." In *Promotio Iustitiae*, edited by Patxi Álvarez SJ, 2–67. Rome: Social Justice Secretariat at the General Curia of the Society of Jesus, 2011.

Somerville, Richard C. J. "The Ethics of Climate Change." *Yale Environment 360: Opinion, Analysis, Reporting and Debate* June 3, 2008. https://e360.yale.edu/features/the_ethics_of_climate_change (accessed September 4, 2018).

Sorkin, Andrew Ross. "BlackRock's Message: Contribute to Society, or Risk Losing Our Support." *The New York Times*, January 15, 2018. https://www.nytimes.com/2018/01/15/business/dealbook/blackrock-laurence-fink-letter.html (accessed January 23, 2018).

Spencer, Daniel T. "Evolutionary Literacy: A Prerequisite for Theological Education?" *Worldviews: Global Religions, Culture & Ecology* 11, no. 1 (March 2007) 83–102.

Spencer, Daniel T. *Gay and Gaia: Ethics, Ecology, and the Erotic*. Cleveland: Pilgrim, 1996.

Spivey, Robert A., et al. *Anatomy of the New Testament: A Guide to its Structure and Meaning*. 6th ed. Minneapolis: Fortress, 2007.

Spretnak, Charlene. *The Resurgence of the Real: Body, Nature, and Place in a Hypermodern World*. New York: Routledge, 1999.

———. *States of Grace: The Recovery of Meaning in the Postmodern Age*. San Francisco: Harper, 1991.

Stanford Environmental Law Society, ed. *The Endangered Species Act*. Stanford: Stanford University Press, 2001.

Bibliography

Stauber, John C., and Sheldon Rampton. *Toxic Sludge is Good for You: Lies, Damn Lies, and the Public Relations Industry*. Monroe: Common Courage, 1995.

Steiner, Gary. "Descartes, Christianity, and Contemporary Speciesism." In *A Communion of Subjects: Animals in Religion, Science, and Ethics*, edited by Paul Waldau and Kimberley Patton, 117–31. New York: Columbia University Press, 2006.

Stern, Nicholas H. *The Economics of Climate Change: The Stern Review*. Cambridge: Cambridge University Press, 2007.

Stone, Christopher D. *Should Trees Have Standing? Toward Legal Rights for Natural Objects*. Los Altos, CA: W. Kaufmann, 1973.

Stone, Christopher D. "Common but Differentiated Responsibilities in International Law." *American Journal of International Law* 98, no. 2 (2004) 276–301.

Stratton, Leeanne, et al. "The Persistent Problem of Malaria: Addressing the Fundamental Causes of a Global Killer." *Social Science & Medicine* 67, no. 5 (2008) 854–62.

Sukhdev, Pavan, and Patrick ten Brink, eds. *TEEB: The Economics of Ecosystems & Biodiversity for International and National Policy Makers 2009 (Executive Summary)*. UNEP: United Nations Environment Programme, 2009.

Suzuki, David T., and Amanda McConnell. *The Sacred Balance: Rediscovering our Place in Nature*. Amherst: Prometheus, 1998.

Swearer, Donald K., and Susan Lloyd McGarry, eds. *Ecologies of Human Flourishing*. Cambridge, MA: Harvard University Press, 2011.

Swimme, Brian, and Thomas Berry. *The Universe Story: From the Primordial Flaring Forth to the Ecozoic Era—a Celebration of the Unfolding of the Cosmos*. San Francisco: Harper, 1992.

Tallis, Heather, and Jane Lubchenco. "Working Together: A Call for Inclusive Conservation." *Nature* 515, no. 7525 (2014) 27–28.

Tanner, Kathryn. "Is Capitalism a Belief System?" *Anglican Theological Review* 92, no. 4 (2010) 617–35.

Haneke, Tom. *John Muir in the New World*. DVD. Directed by Catherine Tatge et al. West Long Branch: Kultur, 2011.

Tauli-Corpuz, Victoria. "Self-Determination and Sustainable Development: Two Sides of the Same Coin." In *Reclaiming Balance: Indigenous Peoples, Conflict Resolution & Sustainable Development*, edited by Victoria Tauli-Corpuz and Joji Cariño, 3–74. Baguio City, Philippines: Tebtebba Foundation, 2004.

Taylor, Bron Raymond, et al. *Dark Green Religion: Nature Spirituality and the Planetary Future*. Berkeley: University of California Press, 2010.

———. *The Encyclopedia of Religion and Nature*. London: Continuum, 2005.

———. "Facebook [Group]." Posted July 31, 2013. https://www.facebook.com/groups/ISSRNC (accessed August 1, 2013).

Taylor, Charles. *Sources of the Self: The Making of the Modern Identity*. Cambridge, MA: Harvard University Press, 1989.

Taylor, Sarah McFarland. *Green Sisters: A Spiritual Ecology*. Cambridge, MA: Harvard University Press, 2007.

Tebaldi, Claudia, and Pierre Friedlingstein. "Delayed Detection of Climate Mitigation Benefits due to Climate Inertia and Variability." *Proceedings of the National Academy of Sciences of the United States of America* (published ahead of print October 2, 2013).

Teilhard de Chardin, Pierre. *The Future of Man* [Avenir de l'homme]. Translated by Norman Denny. New York: Harper & Row, 1964.

BIBLIOGRAPHY

———. *The Divine Milieu: An Essay on the Interior Life*. Translated by Bernard Wall. New York: Harper, 1960.

———. *The Phenomenon of Man*. Translated by Bernard Wall. New York: Harper, 1959.

Tennyson, Alfred Lord. *In Memoriam (A. H. H.)*. London: 1850.

Terhaar, Terry Louise. "Evolutionary Advantages of Intense Spiritual Experience in Nature." *Journal for the Study of Religion, Nature & Culture* 3, no. 3 (September 2009) 303–39.

Terrace, Herbert S., and Janet Metcalfe, eds. *The Missing Link in Cognition: Origins of Self-Reflective Consciousness*. Oxford: Oxford University Press, 2005.

The Earth Charter Initiative. "*The Earth Charter*." The Earth Charter Commission. http://www.earthcharterinaction.org/content/pages/Read-the-Charter.html (accessed September 4, 2018).

The Norwegian Refugee Council. *Climate Changed: People Displaced*. Norway: Norwegian Ministry of Foreign Affairs, 2009.

The White House. *The President's Climate Action Plan*. Washington, DC: Executive Office of the President, June 2013.

———. *U.S.-China Joint Announcement on Climate Change*. Washington, DC: The Executive Office of the President, November 2014.

The World Bank. *Turn Down the Heat: Why a 4⬚ Warmer World must be Avoided*. Washington, DC: International Bank for Reconstruction and Development / The World Bank, 2012.

———. *World Development Report 1999/2000*. New York: Oxford University Press, 1999.

Thistlethwaite, Susan Brooks. *Dreaming of Eden: American Religion and Politics in a Wired World*. New York: Palgrave Macmillan, 2010.

Thomas, Dylan. "Do Not Go Gentle Into That Good Night." *Botteghe Oscure* 8 (1951) 208–10.

Thompson, Allen, and Jeremy Bendik-Keymer, eds. *Ethical Adaptation to Climate Change: Human Virtues of the Future*. Cambridge, MA: MIT Press, 2012.

Tilley, Maureen A., and Susan A. Ross, eds. *Broken and Whole: Essays on Religion and the Body*. Lanham: University Press of America, 1995.

Tillich, Paul J. *Systematic Theology*. Vol. 1. Chicago: University of Chicago Press, 1951.

Tomlinson, Sherrie A. "No New Orleanians Left Behind: An Examination of the Disparate Impact of Hurricane Katrina on Minorities." *Connecticut Law Review* 38, no. 5 (2006) 1153–1188.

Townsend, Peter. "Measuring Poverty." *The British Journal of Sociology* 5, no. 2 (1954) 130–37.

Tracy, David. *Blessed Rage for Order: The New Pluralism in Theology*. New York: Seabury, 1975.

Troeltsch, Ernst. *Protestantism and Progress: The Significance of Protestantism for the Rise of the Modern World*. Philadelphia: Fortress, 1986.

Tucker, Mary E. "Biography of Thomas Berry." The Thomas Berry Foundation. www.ThomasBerry.org (accessed October, 2014).

———. *Worldly Wonder: Religions Enter their Ecological Phase*. Chicago: Open Court, 2003.

Tucker, Mary Evelyn, and John Grim. *Worldviews and Ecology: Religion, Philosophy, and the Environment*. Maryknoll: Orbis, 1994.

Tucker, Mary Evelyn. "World Religions, the Earth Charter, and Sustainability." *World Views: Environment, Culture, Religion* 12, no. 2/3 (2008) 115–28.

Bibliography

Turner, R. Kerry, Stavros G. Georgiou, and Brendan Fisher. *Valuing Ecosystem Services: The Case of Multi-Functional Wetlands.* London: Earthscan, 2008.

UN Conference on Environment and Development. *Framework Convention on Climate Change* 31 I.L.M. 849, 1992.

UNDP. "Fighting Climate Change: Human Solidarity in a Divided World." In *Human Development Report 2007/2008*, 1–18. New York: Palgrave Macmillan, 2007. http://hdr.undp.org/sites/default/files/reports/268/hdr_20072008_en_complete.pdf

———. *The Human Development Report 2010—20th Anniversary Edition.* New York: Palgrave Macmillan, 2010. http://hdr.undp.org/sites/default/files/reports/270/hdr_2010_en_complete_reprint.pdf

UNFPA. "Facing a Changing World: Women, Population and Climate." In *State of World Population 2009* United Nations Population Fund, 2009. https://www.unfpa.org/publications/state-world-population-2009

UN General Assembly. *Convention Related to the Status of Refugees.* United Nations Conference of Plenipotentiaries on the Status of Refugees and Stateless Persons: Adopted 28 July 1951.

———. *Protocol Relating to the Status of Refugees* 31 January 1967, entry into force 4 October 1967.

———. "Report of the United Nations Conference on Environment and Development (Rio De Janeiro, 3–14 June 1992)." United Nations. http://www.un.org/documents/ga/conf151/aconf15126-1annex1.htm (accessed September 4, 2018).

———. "United Nations Framework Convention on Climate Change." United Nations. http://unfccc.int/resource/docs/convkp/conveng.pdf (accessed September 4, 2018).

UNHCR. *Handbook on Procedures and Criteria for Determining Refugee Status Under the 1951 Convention and the 1967 Protocol Relating to the Status of Refugees.* Geneva: Office of the United Nations High Commissioner for Refugees, 1992, 1979.

UN High Commissioner for Refugees. *Briefing Note: The Management of Humanitarian Emergencies Caused by Extreme Climate Events.* Published Electronically: UN High Commissioner for Refugees, April 2009.

———. *Climate Change, Natural Disasters and Human Displacement: A UNHCR Perspective.* Published Electronically: The UN High Commissioner for Refugees, 2009.

Union of Concerned Scientists. "World Scientists' Warning to Humanity (1992)." Union of Concerned Scientists. http://fore.research.yale.edu/publications/statements/union/ (accessed September 4, 2018).

United Nations, Department of Economic and Social Affairs, Population Division. *World Population Prospects: The 2012 Revision, Key Findings and Advance Tables.* New York: United Nations, 2013.

United Nations Population Fund. "What does UNFPA Stand for?" United Nations Population Fund. http://www.unfpa.org/public/about/faqs#acronym (accessed September 4, 2018).

United Nations Statistic Division. "Composition of Macro Geographical (Continental) Regions, Geographical Sub-Regions, and Selected Economic and Other Groupings (Footnote C)." United Nations Statistic Division. http://unstats.un.org/unsd/methods/m49/m49regin.htm#ftnc (accessed September 4, 2018).

United States Conference of Catholic Bishops. *Global Climate Change: A Plea for Dialogue, Prudence and the Common Good.* Washington, DC: United States Conference of Catholic Bishops, 2001.

BIBLIOGRAPHY

———. "Catholic Social Teaching and Environmental Ethics." In *Renewing the Face of the Earth*. Washington, DC: United States Conference of Catholic Bishops, 1994. http://www.usccb.org/issues-and-action/human-life-and-dignity/environment/renewing-the-earth.cfm

US Environmental Protection Agency. "Glossary—Total Maximum Daily Loads." US Environmental Protection Agency. http://water.epa.gov/lawsregs/lawsguidance/cwa/tmdl/glossary.cfm (accessed September 4, 2018).

———. "Waste and Cleanup Risk Assessment Glossary." US Environmental Protection Agency. http://www.epa.gov/oswer/riskassessment/glossary.htm (accessed December 18, 2012).

US Fish & Wildlife Service. "ESA Basics: 40 Years of Conserving Endangered Species." US Fish & Wildlife Service, Endangered Species Program. http://www.fws.gov/endangered/esa-library/pdf/ESA_basics.pdf (accessed September 4, 2018).

US Geological Survey. "Water Science Glossary of Terms." US Geological Survey. http://ga.water.usgs.gov/edu/dictionary.html (accessed December 18, 2012).

Vandentorren, S., et al. "August 2003 Heat Wave in France: Risk Factors for Death of Elderly People Living at Home." *European Journal of Public Health* 16, no. 6 (2006) 583–91.

Veit-Wilson, J. H. "Paradigms of Poverty: A Rehabilitation of B. S. Rowntree." *Journal of Social Policy* 15, no. 01 (1986) 69–99.

Vidal, John. "Bolivia Enshrines Natural World's Rights with Equal Status for Mother Earth." *The Guardian*, April 10, 2011. https://www.theguardian.com/environment/2011/apr/10/bolivia-enshrines-natural-worlds-rights (accessed September 4, 2018).

Vidyasagar D. "Global Notes: Counting the World's Poor—how do we Define Poverty?" *Journal of Perinatology: Official Journal of the California Perinatal Association* 26, no. 6 (2006) 325–27.

Vince, Gaia. "Coping with Climate Change: Which Societies Will do Best?" *Yale Environment 360: Opinion, Analysis, Reporting and Debate* November 2, 2009. https://e360.yale.edu/features/coping_with_climate_change_which_societies_will_do_best (accessed September 4, 2018).

Waks, Leonard J. "Environmental Claims and Citizen Rights." *Environmental Ethics* 18, no. 2 (1996) 133–48.

Walker, Cam. "A Gathering Storm: Climate Change and Environmental Refugees." *Arena Magazine*, December 1, 2003. https://www.highbeam.com/doc/1G1-111895821.html (accessed September 4, 2018).

Walzer, Michael. *Exodus and Revolution*. New York: Basic, 1985.

Wanliss, James. *Resisting the Green Dragon: Dominion, Not Death*. Burke: Cornwall Alliance for the Stewardship of Creation, 2010.

Ward, Peter Douglas. *Under a Green Sky: Global Warming, the Mass Extinctions of the Past, and What They Mean for our Future*. New York: Smithsonian/Collins, 2007.

Warner, Koko, et al. *In Search of Shelter: Mapping the Effects of Climate Change on Human Migration and Displacement*. Bonn: CARE Deutschland-Luxemburg, May 2009.

Washington, Haydn, and John Cook. *Climate Change Denial: Heads in the Sand*. London: Earthscan, 2010.

Watson, Richard A. "Self-Consciousness and the Rights of Nonhuman Animals and Nature." *Environmental Ethics* 1, no. 2 (1979) 99–129.

Weber, Max. *The Protestant Ethic and the Spirit of Capitalism*. Translated by Talcott Parsons. New York: Routledge Classics, 2001 (1904/5, 1930).

Bibliography

Wesley, John. *The Works of John Wesley: The Methodist Societies—History, Nature, and Design*. Edited by Rupert Davies. Vol. 9. Nashville: Abingdon, 1989.

West, Traci C. *Disruptive Christian Ethics: When Racism and Women's Lives Matter*. 1st ed. Louisville: Westminster John Knox, 2006.

Westra, Laura. *Living in Integrity: A Global Ethic to Restore a Fragmented Earth*. Lanham: Rowman & Littlefield, 1998.

Westra, Laura and Mirian Vilela, eds. *The Earth Charter, Ecological Integrity and Social Movements*. London: Earthscan, 2014.

"What are B Corps?" *Certified B Corporations*, https://www.bcorporation.net/what-are-b-corps (accessed January 3, 2018).

White, Lynn, Jr. "The Historical Roots of our Ecological Crisis." *Science* 155 (1967) 1203–1207.

Whitman S, et al. "Mortality in Chicago Attributed to the July 1995 Heat Wave." *American Journal of Public Health* 87, no. 9 (1997) 1515–1518.

Wildiers, N. Max. *The Theologian and His Universe: Theology and Cosmology from the Middle Ages to the Present*. New York: Seabury, 1982.

Williams, David R. *Wilderness Lost: The Religious Origins of the American Mind*. Cranbury: Susquehanna University Press, 1987.

Williams, Stephen E., et al. *Towards an Integrated Framework for Assessing the Vulnerability of Species to Climate Change* 6, 2008.

Wilson, David Sloan. *Darwin's Cathedral: Evolution, Religion, and the Nature of Society*. Chicago: University of Chicago Press, 2002a.

Wilson, Edward O. *The Creation: An Appeal to Save Life on Earth*. New York: W. W. Norton & Company, 2006.

———. *The Diversity of Life*. Cambridge, MA: Belknap, 1992.

———. *The Future of Life*. 1st ed. New York: Alfred A. Knopf, 2002.

World Health Organization. "WHO Position Paper on Poverty and Ill Health." *World Health Organization Regional Office for Africa* (October 1999).

Worster, Donald. *A Passion for Nature: The Life of John Muir*. Oxford: Oxford University Press, 2008.

Wright, Ronald. *A Short History of Progress*. New York: Carroll & Graf, 2004.

Wulf, Andrea. *The Invention of Nature: Alexander von Humboldt's New World*. New York: Knopf, 2015.

Wybrow, Cameron. *The Bible, Baconianism, and Mastery Over Nature: The Old Testament and its Modern Misreading*. New York: Peter Lang, 1991.

Zalasiewicz, Jan, et al. "Are we Now Living in the Anthropocene?" *GSA Today: A Publication of the Geological Society of America* 18, no. 2 (2008) 4–8.

Index

actions, individual as cumulative, 102n55
"active partner," metaphors for the Earth as, 82
African continent
    droughts and desertification plaguing, 37
    low emissions relative to the rest of the world, 39
    over a quarter of the world's poor, 30
    water already a precious commodity, 40
agri-industries, 88
American Teilhard Association, 72
anthropocentric focus, 75, 96
anti-Pelagian writings, of Augustine, 52
Aquinas, Thomas. *See* Thomas Aquinas
Arianism, 54
Aristotle, 19n56
Augustine of Hippo, 47, 52–55, 65, 75
awe and wonder, recapturing, 106–7

Bacon, Francis, 8–9
Bangladesh, xix, 23, 31n16
Barbut, Monique, 92n42
Bauman, Whitney, 85n28
Bavington, Dean, 13n38
Bell, Daniel, 12n36
Bengal tiger, losing critical habitat, 36
Benzoni, Francisco, 59n42
Berger, Peter L., 13n40
Berry, Thomas, 63n57, 71–75, 71n85, 74n98, 90

biblical creation stories, themes contained in, 47–49
biblical stories, Christians drawing on ancient, 96–97
Bieringer, Reimund, 90n37–91n37, 91
birds, beasts, and human beings, all drawing a breath of life, 51–52
BlackRock, 89n34
"blessed to be a blessing" banner, xix
bodies
    corruptibility of, 52, 52n19
    needs of as satiable, 97
    requiring access to clean air, water, and food, 94–95
body theology, 70
Boff, Clodovis, 21
Boff, Leonardo, 19
Bolivian government, recognized the rights of *Pachamama* (Mother Earth), 80–81
bridging the gap, between what is and what ought to be, 14, 16
Brown, Lester, 85, 93n43
butterfly, on the summit of a mountain in China, xviii

Calvin, John, 10
capacity building programs, challenges of community-based, 33–34
carbon dioxide, impact of, 38n33
carbon emissions, 37, 38
carbon footprint, 39, 89
Cavanaugh, William T., 64–65

Index

certified benefit corporations ("B Corps"), 88–89
change, seeds of, 101–4
chaos, as the opposite of a sacred cosmos, 13n40
cholera, 41, 41n44
Christ, dwelling in and amidst the world, 24
Christian clergy, role of, 4
Christian communities, 94, 95n47–96n47, 99n50, 102
Christian deviations, Augustine attacking, 54
Christian faith. *See also* faith
　author's early exposure to, xvi
　expanding this garden of, 100
　reconstructing, 69n79
Christian ideas, 62–63
Christian moral life, as attentive to what is going on in the world, 14
Christian Scriptures, beginning with a story about the soil, 96
Christian soteriology, ecologizing, 69n79
Christian theology, rituals, and traditions, growing in response to a life lived in faith, 62
Christian thinking, skepticism regarding the potential effectiveness of greening, 5n11
Christian traditions
　creation-centered, 16
　rituals in, 62, 64
　role of, 4, 45–46
　shaped by stories, 7
Christianity, as the most anthropocentric religion, 18–19
Christians
　acknowledging our humanity is bound up with all other life on this planet, 18
　called to be responsive and responsible, 14
　drawing inspiration from natural processes, 73
　ethic of eco-justice, 90
　imagining God as the vivifying power for all life, 55–56
　including needs and concerns of the most vulnerable, 18
　longing to really see things as they are, 20
　needing new words about God, 99–100
　North American terrified by the poor, 65n66
　pulling forward prominent streams of pre-modern Christian thought, 59
　rejecting a vision of the Earth as a place to be dominated, subdued, and managed, 51
　reporting lower levels of environmental concern, 5n11
church
　author's encounters with, xvi
　needing more words about God, 100
citizens and consumers, taking action collectively, 86
civilizations
　requiring a major paradigm shift, 75
　unitive vision for, 72
Clark, Gregory, 11n33
climate accords, negotiating, 103n59
climate change
　accepting or denying, 3
　borne by the poorest and most vulnerable people, 26
　as both a moral issue and a social justice issue, 27
　environmental impact of, 103n59
　as global and local, 37–43
　policy, 1
　presenting as a serious problem, 2
　resulting from carbon-heavy industries, 85
　US public's general lack of concern for, 5
climate denial, confronting, 1, 2
climate resiliency, 25
climate scientists, 2, 6
climate vulnerabilities, for sensitive populations, 25
climate-induced displacement, social impact of, 103n59

Index

coal-fired power plant, building in the Sundarbans, 36
commitment, to all life at the margins, 16–18
common good, 47, 56–62, 61n50, 86
communion table, 63–65
communities
    Christian, 94, 95n47–96n47, 99n50, 102
    indigenous, 74n99, 78–79
    of influence, 6
    of mutual responsibility, 97
    as stakeholders of a company, 89
    of suffering people, 24–25
companies, not charged for spewing waste products, 84
consciousness, autonomous as wholly illusory, 67n69
conspicuous consumerism, shift toward, 12n36
conspicuous consumption, 11, 11n32
consumer products, global appetite for, 83
consumer-driven demand, power of, 88
consumerism, as akin to a "global religion," 12n37
Corporate Social Responsibility (CSR) officers, companies adding, 89
cosmic Christ, 70–71
cosmic journey of life, beauty and grandeur of, 107
cosmos, 8, 13n40, 52–56, 62
creation
    inherent goodness of, 52
    order and diversity of creatures, 58
    as the purview of theology, 7
    story informed by modern science, 73
    waning of a strong doctrine of, 8n19
    worth of, 97–98
creation-centered worldviews, recovery of premodern, 69n79
"creaturely theology," 73n96
creatures, many and diverse, 62
cultural polarization, among members of the public, 3
cultural power of religion, 63n56

cultures of encounter, striving to create, 91

Dahle, Øystein, 85
Dalai Lama, 93n45
dandelion story, 106
De La Torre, Miguel A., 17
death, more likely for impoverished women, 31
deChant, Dell, 96
DeCrane, Susanne M., 61, 61n48
degradation, 21, 26, 101
delta regions, home to large population centers, 41
Descartes, René, 9
despair, 24
"developed" and "developing" countries, 29n10
developed countries, offshoring manufacturing, 33
developed world, overconsumption by, 91
development projects, xviii
disaster-prone areas, options about abandoning, 38
disparities, surrounding climate change, 39
distracted life, contrasted with the examined life, 95
"dominion" passages, interpreting christologically, 50n11
Donatism, 54
Douglas, William O., 80
dualistic images of God, 66, 68
dualistic metaphysics, 8
Durning, Alan Thein, 11n32

Earth
    as a deteriorating body in desperate need, 70
    ecological limits of, 98
    reducing to a collection of objects, 51
Earth Charter, 103–5
Earth Charter Commission, 103
Earth community, becoming responsible members of, 15

Index

earthen clay bodies, human persons as, 47–52
earthy sacramentalism, 47, 62–66
ecocentric worldview, 74n99
eco-justice, ethic of, 90
ecological and climate crisis, Christian traditions partly complicit, 19
ecological and evolutionary Earth sciences, recognizing a role for new spiritual-religious creation stories, 99
ecological beings, imagining ourselves as, 67
ecological costs, neglecting, 85
ecological crisis, 6, 19
ecological degradation, consequences of, 26
"ecological footprint" calculators, 83n22
ecological humility, 96
ecological image, 57
ecological problems, 34
ecological responsibility, 36, 90
"ecological self," 67
ecological systems, xx, 13
ecological theology, 70n82
economic calculations, 83
economic development and ecological conservation, necessity of both, 36
economic opportunities, accessing, 95
economic success, as a sign of God's favor, 12
economically disadvantaged communities, unjust suffering experienced by, 86
"The Economics of Ecosystems and Biodiversity" (TEEB), 87n31
economy, private, tending to oversupply goods, 85
ecosystem-dependent livelihoods, 25
ecosystems
    with legally recognized standing, 80
    ratio of benefits to costs for protecting, 87n31
"Ecozoic Era," 72
Ecuadorian people, giving nature rights, 80
electrical power generation, in places like Bangladesh, 35
"electricity haves," and "have-nots," 39
Elsbernd, Mary, 90–91, 90n37–91n37
empiricism, 15n46
Endangered Species Act (ESA) of 1973, 77, 77n1
energy, 83, 84, 88
enfleshed spirituality, 47, 66–71
Enlightenment legacy, requiring re-evaluation, 15n46
environmental and social challenges, not receiving adequate attention from religious traditions, 105
environmental community, attempts to improve scientific literacy, 2
environmental degradation, overwhelming poverty and, xx
environmental objects, conferral of standing upon, 80
equity, interconnection with sustainability, 16, 92
"erotic ethic of eco-justice," 99n50
ethic, of eco-justice, 90
ethical arguments, pushing for action on climate change, 101n54
ethics
    appropriate questions in, 101n53
    integrating with a foundation of facts, 101
    religious, 14
European Enlightenment, influence on dominant streams of Western thought, 15n46
evangelical Christians, 5, 5n10
Evangelical Lutheran Church in America (ELCA), 33n27
Evans, G. R., 54n26
everyday life, banality of, 20
evolution, telling as an epic story, 73
evolutionary and ecological sciences, Christians making peace with, 104
examined life, making space and time for, 95
Exodus story, of a people's liberation, 14n42
exposure risks, 25
externalities, 85, 85n28

Index

extinction event(s), 35, 94
extreme poverty, defined, 27

factories, people shifting to, 10n27
faith. *See also* Christian faith
  connected with action, xx
  gritty kind of, 96–97
  interconnections with equity and ecology, xv
  living as a people of, 24
  open to the unknown but grounded in living experience, xvi
  people of, retrieving a theology of creation, 98
faith-based religious communities, environmental community not building bridges and coalitions with, 2–3
farmers, 33, 88
Fasching, Darrell J., 96
female-headed households, 32, 32n21
15th Conference of the Parties (COP15), 39
Fink, Laurence D., 89n34
flooding, 41, 41n44
foods, selection of, 88
Pope Francis, 51
French, William C.
  on Aquinas, 57–58, 58n40, 59n44, 68n75
  on Aristotle, 19n56
  on externalities, 85–86
  on the premodern context, 7n16
  on society lacking a story "fitting" for the ecological crisis, 19
fuel, cost of, 84
future, grimly predicted, 6

Gandhi, Mohandas, 13
Ganges river, accorded status of a legal person, 79
Garanzini, Michael J., 18, 18n51
Garvey, James, 101n53
Geertz, Clifford J., 13n39
Genesis, 48n4, 75
Genesis 1, 47–49
Genesis 2, 48, 49, 50, 96

Genesis 2:7, 51
Geocentric Model, of the Ptolemaic system, 84n23
Germany
  recentering economies around renewables, 88
  respect and protection for "animal dignity," 78
Gibson, William E., 90
global commons, spewing waste into, 84
global environmental crisis, 6
"global ethic," in the Earth Charter, 103, 103n62–4n62
"Global Marshall Plan," Gore's call for, 94n46
global missions office, xviii
global poor, living primarily in South Asia and Africa, 28
global poverty, 27, 101
Gnosticism, 54, 55
God
  activity in the world, 67n71
  binding human beings to each other, 75
  as the Creator dwelling in and amidst all things, 69n79
  delighting in this good Earth, 21
  dwelling both in and beyond all things, 69
  enmeshed within the earth's ecosystems, 68–69
  instruction to humanity, 50
  invitation to all people, 91
  light of, 23, 24
  meeting where God dwells, 24
  real presence through bread and wine, 63
  seeking in all things, 107
  speaking creation into being, 48
  there in and amidst the turmoil, violence, and death, 24
  transformative love for the world, 66
  as a vivifying power, 75
God-moments, mountaintop experiences as, 23
"God's call," listening to, 14
good cosmos, 52–56

Index

Gore, Al, 1, 6, 94n46
governance, facilitating good, 92
governments, holding industries and corporations accountable, 36
grand visions, necessary to guide big changes, 72
Great Ape Project, 77
greening
  of Christian soteriology, 69n79
  of Christian thought, 5n10
Guatemalan rainforest, hike through, 44
Gustafson, James, 90n35

Häring, Bernard, 14
Hawa, 22, 24
healthcare, 22
Hiebert, Theodore, 48–50
hierarchical view, in Genesis 1, 49
hierarchical worldview, in Western philosophical and theological thinking, 68n75
Himalayas, increased melting of snow and ice in, 37
Hinduism, Mohandas Gandhi appealed to, 13
historical periods, Berry's framework and naming of, 71n86
"Historical Theory of Giambattista" (Berry), 71n86
history, stories shaping, 6–16
Homo administrator, 13n38
hope, 104–7
HPI-1 (Human Poverty Index), 29
human(s). *See also* human species; humanity
  ability to persevere, 104
  agency of, 10
  basic needs of all, xxi
  connected to every other being, 73
  dominion over the Earth, 47–48
  in an ecological reality, 98
  embeddedness within creation, 15
  as an embodied, ecologically embedded earth creature, 67
  embodied, relational view of, 90
  existing for the good and perfection of the larger universe, 60n47
  finding solace, joy, and beauty amidst unimaginable despair, 23
  framing in a more humble and communitarian framework, 47
  growing population of, 83
  as highly social, 57
  as *Homo administrator* or management species, 12–13
  inspired toward action by stories, 94
  intimacy with the Earth, 82
  joined a family of breathing creatures, 51
  life-spans of, 7n16
  as the locus of rational thought, 10
  positioning in a framework of planetary humility, 50n11
  power and presence on the Earth, 98
  requiring opportunities for leisure, rest, and refreshment, 95
  story as part of a larger story of life on Earth, 73
  suffering from global and social dynamics, 34
  technological appropriation of nature for, 9
  Thomistic understanding of, 60
  understanding within a cosmological frame, 58n40
  vocation as one of dominion and supervision, 49
  welfare tied to planetary well-being, 35
human bodies. *See* bodies
Human Development Report Office, on climate change, 38–39
Human Poverty Index (HPI-1), 29
human species. *See also* human(s); humanity
  capable of mobilizing change, 106
  as a geological force, 42
  on a mote of dust suspended in a sunbeam, 93n44
  perceiving nature as fractured and vulnerable, 87
  persisting in plucking this miracle of life out of existence, 93
human-Earth relations, great work of transforming, 103

140

Index

humanity. *See also* human(s); human species
　ability to survive and thrive as inseparable from the flourishing of the planet as a whole, 71
　on the cusp of a major paradigm shift, 84n23
　ecological embeddedness of, 35
　molded from the Earth and commissioned to till and keep the land, 48
　as a participant in the broader community of the universe, 58n40
　physically reimagining, 76
　relations to other species, 35
　shaped in the "image" or "likeness" of God, 48
hurricanes, 26, 26n5

ideas
　about what it means to be human, 19, 47
　Christian, 62–63
　cultural power of religious, 63
　premodern, 46n2, 82
　religious, 14
image of God, as a trinity of persons, 90n37–91n37
inadequacy, in the face of great despair, xix–xx
incarnation, 24
inclusion, for the outcast, 18
inclusivity, for the marginalized and excluded, 21
An Inconvenient Truth film (Gore), 1
indigenous communities, 74n99, 78–79
indigenous cosmologies, 82n21
individual autonomy, as a primary concept of the self, 15
individualism, Western societies internalizing an anthropology of radical, 66–67
individualistic images of the human person, in Christian theology, 66
individualistic self, 65, 67n71
individuality and autonomous rights, envisioning humanity merely in terms of, 15n46

industrial food systems, transforming, 88
Industrial Revolution, bringing changes, 10
industries, expansion of, 10n27
injustice, 17, 26, 34, 36
institutional processes, of the world, xx
institutions, as arbiters of justice, 36
Integrated Community Development, 33n27
international community, paralyzed by the challenge of climate change, 42
international corporations, benefiting from cheap labor, 33
International Energy Agency, 39
international scientific community, consensus on environmental and climate changes, 38n34
International Society for the Study of Religion, Nature and Culture (ISSRNC), 5n11

Pope John Paul II, 51n16, 62
Johnson, Elizabeth, 16, 46n2, 70n82
journalists, undue attention to climate deniers, 2
justice, 17, 92n41. *See also* social justice
justice as participation, 90, 90n37–91n37, 91, 92, 93

Kant, Immanuel, 10, 72n92
Kantian notion of the world, 72
King, Martin Luther, Jr., 13–14, 103n60
Kuhn, Thomas S., 84n23

labor and wealth creation, in the industrializing West, 10n27
land, as sacred and worth preserving, 81
Lane, John, 103n61
Latin America, recognizing nature's standing, 80–81
Laudato Si' (Pope Francis), 51
legal standing, expansion of, 79
liberation, for the imprisoned, 18

Index

liberation theology. *See also* theology
- on concern for the poor and oppressed, 16, 16n48
- critiquing the selfish, pretentious self, 67
- expansion of, 18

life
- bearing witness to mutual dependency, 98
- on Earth, 35
- financial benefits of preserving and protecting, 87n31
- fragility of, 18
- of justice as participation, 91

life at the margins
- commitment to all, 16–18
- God loving and cherishing all forms of, 24

"Living Cosmology: Christian Responses to Journey of the Universe" conference on Berry's work, 71n85

lo cotidiano (experiences of everyday life), 99

local impacts, of climate change, 37

Luther, Martin, 10, 64

Lutheran Health Care Bangladesh (LHCB), 33n27

MacGillis, Mariam, 100
machine metaphor, 8, 9, 12–13
malaria, 40–41
mandate "to till" the earth, more appropriately translated as "to serve," 50
Mander, Jerry, 82n19
Manichaeism, 54, 55
Māori tribe, 78–79
market systems
- affecting transformational change, 88
- needing to become more sensitive to changing needs, 87
- reshaping to preserve planetary ecosystems, 83
- responsible accounting as the foundation of, 85n28
- rise of reliant upon the transformation of natural resources, 9
- taking a full account of costs, 86

Marx, Karl, 61n50–62n50
mass consumption, 11n32
mass extinction, 98
material world, 52
matter and body, as intrinsically evil, 55
Maya, 45
Mayan temple complex, 44
McFague, Sallie, 15, 66–70, 69n79, 69n80
medical care
- required for human bodies, 95
- women and girls lacking access to basic, 22

men
- not finding work, 33
- planning to leave their village, 34

men and boys, valuing more than girls, 22–23

mental and emotional support services, access to professional, 95

Merchant, Carolyn, 8, 81–82
Mesoamerican pyramid, Temple IV, sunrise from atop, 44–45
metaphysical dualism, dispute on, 54
microbial contaminants, difficult to discern in water, 42
Middle Ages, Black Death characteristic of the late, 57
Miles, Margaret, 64n59
military and economic power, elite acting as a "stick," 85n28
mind/body metaphor, God as orderer and controller of the universe, 69n80
minds, requiring access to educational opportunities, 95
ministers, skilled as storytellers and cultural translators, 4
"missing women," in places like Bangladesh, 22
mobilization, theology of, 90–101
model, for creating a "safe and just space for humanity," 102–3
moderate poverty, defined, 27

Index

moral arc of the universe, bending toward justice, 103, 103n60
moral boundaries, 11
moral concern, for rights held by non-human beings, 78
moral imagination, 79–83
moral life, Christian, 14
moral responsibility, self-conscious awareness including, 68
moral worth, expansion of, 79
mountaintop experiences, 23
Multidimensional Poverty Index (MPI), 29–30
myth-making, on an epic scale, 72

Nash, James, 9, 19
Nathan, William. *See* Berry, Thomas
National Oceanic and Atmospheric Administration (NOAA), 77n1
national safety net, 26
nations, affluent lacking a sense of urgency, 40
natural disasters
　as a matter of life and death, 7n16
　women more likely to die in, 31
natural resources
　exploitation of, 10n27, 60
　protecting and conserving, 87n31
　solutions addressing the disproportionate consumption of, 84
natural world, God's all-pervading presence in, 63
nature
　as either for or against the rest of, 68
　inability to accommodate the status quo, 87
　as inherently good in pre-modern theology, 8
　as a kind of machine, 9
　leaving space, 91n40
　more broadly construed rights for, 78
　ongoing debate over rights for, 78n6
　regarding as imbued with inherent rights, 79n10
　seemed limitless in power and expense, 7n16
voiding a healthy sense of its moral worth, 59n42
Nelson, James E., 70
networks, coverage of controversial issues, 2
"new creation story," of Berry, 72
new stories
　creating, 104
　critically and creatively imagining, 74–75
Newton, Isaac, 8–9
Niebuhr, H. Richard, 14
Nobel Peace Prize, 1
normativity of the future, 92n41
North Park University, author flourishing at, xvii

Oelschlaeger, Max, 12, 64n58, 96
openness, to new responses and new insights, 101
orientation, calls for a shift in, 6n15–7n15
overconsumption, by the world's wealthiest, 83, 84

Pachamama (Mother Earth), 81
panentheistic view, of God and world, 69
paradigm shift, 83–89, 84n23
participatory relationships, 91
"partnership ethic," 82
peaceable kingdom of God, 37, 92n41
peanut butter, author's longing for in China, xviii
Pelagianism, 54
peonies, qualities of, 20
people. *See* human(s)
perspectives, soliciting diverse, 99
Pixley, Jorge V., 21
planetary degradation, 26, 101
planetary salvation, replacing concern for personal salvation with, 66
planetary suffering, various forms of, 24
plate tectonics and continental drift, theory of, 84n23
Platonic dualisms, 52, 55

Index

Platonic theory of body and soul, rejected by Augustine, 55n35
point-source pollution, 37–38, 38n33
pollutive industries, reduced subsidies for, 88
poor and oppressed
  concern for, 16
  exposed to some of the most devastating diseases, 40
  naming and creating space for, 17
  suffering disproportionately from environmental degradation, 21, 30
populations, largest as increasingly urban, 87
Porter, Jean, 61n50
postmodernism, 67
poverty
  author's encounters with, xvi, xix–xx
  categories of, 27
  as a cruel injustice, 21
  defining by income, 28, 29
  exacerbated by social context, 32
  global, 27, 101
  international concentrations of, 30
  multifaceted complexity of, 29
  secondary, 31
premodern ideas, 46n2, 82
premodern stories, 74n99
pre-modern West, creation as the purview of theology, 7
premodern worldviews
  not all alike, 81n18–82n18
  retrieval of ecocentric aspects of some, 74n99
"Priestly" accounts, of creation, 48n4
private property, 59, 60, 60n46
privilege and power, application of, 86
problems, acknowledging the breadth and depth of, 14
production, better accounting for the full costs of, 86
production and consumption, costs of not fully accounted for, 84
property
  collective use of during the modern era, 60
  owning, 74n98
  private, 59, 60, 60n46

Protestantism, rise of, 11n33
public transportation projects, 92n42
pursuit of wealth and its generation, as an end in itself, 12

radical theology, as a call to ongoing transformation, 65–66
rainforest, appeared to miraculously come to life, 45
Rasmussen, Larry, 6, 9n21, 11n33, 49n6, 50n11
rationality, humanity sharing only with angels, 58n40
Raworth, Kate, 102n56
reductionism, 15n46
regard for the "self," as wholly illusory, 67n69
relationships, broken among and between people and the planet, xxi
relative poverty, 28
"relevant whole," relating to understanding the human person, 90n35
religion(s)
  acting to construct a cosmos of meaning, 13n40
  author's personal experiences with, xv–xvi
  diversity of, 16
  inspired Christian commitment to personal industry and frugality, 10
  inspiring and facilitating the Earth Charter, 104n63
  as "a system of symbols," 13n39
religious and spiritual narratives, shaping evidence-based conclusions of scientists, 4
religious stories
  helping people face incredibly daunting hardships, 42–43
  rallying to current challenges, 14
  shaping the way people live in the world, 4
religious symbols, 13, 14
religious traditions, 15, 64n58
remittances from abroad, 33n25
resiliency, 25, 26

Index

right ordering, seeking God in solidarity with others, 55
rights-based approach, versus a value-based approach, 78n7
rituals, in Christian traditions, 62, 63, 64
Rockefeller, Steven C., 103n61
Rolston, Holmes, III, 91n40
Roman Catholic social theory, concept of subsidiarity in, 102n55
Ruether, Rosemary Radford, 64n58
rugged individual self, overcoming sinfulness through hard work and sheer determination, 12

Sachs, Jeffrey D., 27–28, 85
sacred quality, of the cosmos, 13n40
Safe and Just Space (Raworth), 102n56
safety net, facing a near-constant onslaught, 26
Sagan, Carl, 93n44
salt-water contamination, of drinking water, 41–42
salvation
  rescuing the divine spark, 55
  with the world, 71
Santmire, H. Paul, 57n38
Schaefer, Jame, 8, 53, 53n21
Scheid, Daniel P., 58, 58n41
science
  insights of, 97–98
  not easily integrated into deeply set stories in most people's minds, 5–6
  rise of modern in Western Europe, 8
scientific information, filtered, 3
scientific inquiry, not making ethical normative claims, 3n5
scientific knowledge, 2, 3, 84n23
scientific method, 82
scientific philosophies, rise of overly reductionist modern, 10
Scientific Revolution, 81, 82
Scriptures, 51, 96, 97
secondary poverty, 31
seeds, of change, 101–4

self, 12, 45–46, 65, 67, 67n71
self-conscious awareness, of human beings, 68
self-image, 66
self-reflection, pilgrimage of, xvii–xviii
self-worth, not coming from material acquisition, 97
sensitive populations, 25, 30
servitude, of slave to master, 50
"shared being," sense of, 93n45
shared common spaces, exploitation of, 60
short-term needs, prioritizing, 35
Sierra Club, 1, 4, 79–80
sin, 52–53
smart development policy, facilitating, 92
Smith, Adam, 11n33
"so what?" question, evaluating ethical thrust, 101
social and ecological concerns, as connected, 34–37
social and political movements, in Latin America, 81
social issues, connections between, xx
social justice, 36, 90, 101. *See also* justice
social nature, of humanity, 60, 61n48
social safety net, 28
social structures, xx, 37
social support systems, unavailable around much of the world, 25–26
social systems, 35, 89
societies, as arbiters of justice, 36
soil, human(s) made from and for, 96
soteriology, modern predominance in Christian theology, 69n9
South Asia
  flooding and waterborne diseases, 37
  more than half of the world's poor living in, 30
spaces
  creating, 98–100
  promoting for each person, 92
Spain, protection of great apes, 78
spark of the divine life, in human bodies, 55

Index

species
  all connected through a shared evolutionary story, 67
  creating legal space for the needs of other, 77n2
  distinctions between, 68
  legal protections for other, 77
Spencer, Daniel, 99n50
spiritual practices, cultivation of world-aware, 100
spirituality, emerging from a body theology, 66
Stations of the Earth, at Genesis Farm, 100
stockholders, living far from pollutive externalities, 86
Stockholm Environment Institute, 102n56
stories
  clergy telling powerful, 4
  needing new, 1–21
  needing to arouse our love and respect for the Earth's own inherent worth, 98
  needing to link faith with action, 96
  old, not incorporating images of the world, 74
  shaping history, 6–16
  that matter, 71–76
  weighing more on hearts and minds than argument alone, 106
storytelling
  creative potential not yet fully tapped on climate change, 3
  from vastly different time periods, 49
"storytelling culture-dwellers," human beings as, 96
Structure of Scientific Revolutions (Kuhn), 84n23
subsidiarity, in models of governance, 102n55
subsistence farmer, image of, 50
Sundarbans ecosystems, degradation of, 36n29
sunrises, as opportunities for profound transitions, 45

Superstorm Sandy, impact on homeowners, 26, 26n4, 26n6
"sustainability," as a social revolution, 84n23
sustainable development, 36, 92
Swimme, Brian, 73
Switzerland, recognizing animals as "beings," 77–78
system, profit for so few at the expense of so many, 37
systemic aggression, against the earth, 19

Taylor, Bron, 5n11
Taylor, Charles, 53n20, 56n36
theological activity, to study the world as it is, 24
theological inquiry, people, God, and nature as spheres of, 7
"theological naturalism," 57n38
theological reflection
  grounding by asking So what? 101
  profound dualism influenced trajectories of, 8
theology. *See also* liberation theology; theology of mobilization
  body, 70
  Christian, 62
  as the clarification of the God-human relationship, 10
  cleaving of the world of nature from, 10
  creaturely, 73n96
  drawing on the best science of the day to inform, 58
  ecological, 70n82
  left with the spheres of God and people, 8
  meaning simply "words about God," 98–99
  radical as a call to ongoing transformation, 65–66
theology of creation, retrieving, 98
theology of mobilization, 90–101. *See also* theology
  action as the most important characteristic of, 105
  described, 93

Index

driving, 101
inspiration for, 93n43
mobilizing Christian communities to partner with all others, 94
within Christian communities and across other religious traditions, 102
Thomas Aquinas
  on both the human person and the world having value before God, 62
  on each creature existing for the good of the whole order, 68n75
  on a person not possessing external things as his alone, 59
  on private property, 60n46
  recognizing intrinsic value in "nature," 60n47
  on striving for the common good, 75
  supporting a robust anthropocentrism, 58n40
  teachings of, 56
  theology of the common good, 47
Thomistic theology, 56–59, 61
"throw-away" culture, 83, 97
traditions. *See* Christian traditions; religious traditions; wisdom traditions
trajectory of time, situating us in a community of people, 107
transformation, call to ongoing, 65–66
transformative experiences, shaping the author's life, xv
triple bottom line, of B Corps, 89

uncertainties, approaching with fear and hesitation, 20
United Nations, Homan Poverty Index (HPI-1), 29
United Nations Declaration of Rights for Great Apes, 77
United Nations Population Fund (UNFPA) report, 31n15
United States, population employed in non-ecosystem-dependent livelihoods, 25
United States Fish and Wildlife Service (FWS), 77n1
universal invitation, of God to all people to be in relationship, 91
universe
  as the body of God, 69
  as a collection of objects rather than as a communion of subjects, 72n93
  participating in the divine goodness, 62
urban centers, design of, 88
urban megacities, mass movement of people into, 32
urban prosperity, dreams and visions of, 32
urgency and a shared call to action, driving the need for a theology of mobilization, 101

value-based approach, versus rights-based approach, 78n7
values and beliefs, individuals using knowledge to affirm preexisting, 3
Vatican, referencing Genesis 1 almost exclusively, 51
Vico, Giambattista, 71n86
violence, abuse, and neglect, far more common among the relative poor, 28
vulnerability, as the reverse of resiliency, 25

wastes, safe processing of not always possible, 85
water shortages, 40, 40n40
wealth
  generation of, 10, 11n33
  pursuit of as an end in itself, 12
  resulting in real moral dangers for the Christian, 10
  as a sign of God's favor, 11–12
Weber, Max, 11–12
Wesley, John, 10, 11
Western thought, turn in the general ethos of, 7
Whanganui River, accorded its own legal identity, 79

what is good, recognizing the rights of a meadow and a creek and woodlands, 74
White, Lynn, 18–19
Wildiers, N. Max, 8
wisdom, embodied in the world's religions, 42
wisdom traditions, 13, 46
woman, in Bangladesh needing milk for her baby, xix
women
    lacking access to basic medical care, 22
    left with sole responsibility for their family, 33
    living in the world's most economically impoverished communities, 30–31
    more likely to stay behind in disaster-prone areas, 32
workplace diversity and sensitivity, promoting, 89
world
    ceased being a primary revelation for the word of God, 8
    as a divinely inspirited body, 69n80–70n80
    engagement with, 14, 20–21, 22–43
    as a good and ordered place, 53
    as inherently good, 52
    as an inspirited body, 69
    needing stories big enough, 72
    studying as it is, as a theological activity, 34
worldly desires, distracting Christians from the gospel, 11
worldviews
    emerging hybrid of pre-modern and modern, 81
    of persons, 3
wound, of poverty and wretchedness, 19

Yahwist accounts of creation, in tension with "Priestly" accounts, 48n4
Yamuna river, accorded status of a legal person, 79

Zimbabwean woman, defining poverty, 28–29